J. Chandler

Identity of Light and Nerve Force

J. Chandler

Identity of Light and Nerve Force

ISBN/EAN: 9783337270056

Printed in Europe, USA, Canada, Australia, Japan

Cover: Foto ©berggeist007 / pixelio.de

More available books at **www.hansebooks.com**

IDENTITY

OF

Light and Nerve Force.

BY

J. CHANDLER.

TITUSVILLE, PA.:
GRAHAM & LAKE, PRINTERS.
1879.

Introduction.

Some twenty or thirty years ago, by the mistake of a person who was attending me in a sickness, I took twelve doses of quinine at a single dose.

The medicine threw me into a delirious, half-conscious sleep, and for two or three hours a running flood of panoramic pictures seemed to pass before my eyes, accompanied with humming, ringing and other noises in my ears.

The pictures that flashed before my vision were pictures of events in my past life, one moment accurately and well defined as from memory, but oftener in detached, disconnected parts, disarranged as to time and position, and in a sad jumble.

Some of the noises in my head were as loud as the report of a pistol, then like a band of music, quickly changing to a hum or ringing sound. I had attended a circus and caravan not long previous and circus scenes and show were interjumbled with work, study, plays and griefs, some in mere passing glimpses, others clear and well defined.

For about a week afterward I could reproduce panoramic pictures in my eyes by merely pressing the eye-ball with my finger.

By pinching my ear, or pressing my skull back of my ears I could reproduce sounds in my ears that for a moment seemed as natural as actual sounds.

Organic forces have always been my favorite study and the phenomena of reproducing panoramic pictures in my eyes by the mere pressure of the finger impressed me very strongly with the notion that light and nerve forces were identical.

All the knowledge which I have since acquired of the phenomena of light, and the phenomena of nerve force, have served to confirm this notion.

Some of the facts and phenomena which have served to confirm this notion of their identity are presented in this pamphlet.

Of the three prominent phases of identity I have in this pamphlet dwelt more especially on the phase that is manifest in mental phenomena—instinct, sensation, knowledge, and memory.

The influence of nerve force on muscles and locomotion, and its influence on the propagation and growth of organic life, will be discussed more fully in a future edition in connection with a discussion of the nature of muscular force.

Identity of Light and Nerve Force.

In asserting the absolute identity of light and the force that traverses the nerves of animal life, their identity is to be understood in the same sense as we understand the identity of galvanic and frictional electricity; identical in the same sense in which heat eliminated by combustion and heat held in the dormant state, as in ice water, is identical; identical in the same sense in which brown elastic sulphur is identical with flowers of sulphur; identical in the same sense in which hard, solid ice is identical with fluid water.

In these examples of the same elements and material substance manifesting very different properties, we catch a glimpse of what is understood as allotropic conditions of matters. While in these examples of what are considered different phases of the same force, we obtain also a glimpse of what I have termed allotropic conditions of force. In these conditions we have manifestations of different properties and effects by the same force, and manifestations of different properties by the same substance; these being allotropic connections of force and allotropic conditions of matter. All the facts and phenomena presented by philosophers of the transmutation of one kind of matter into another, or of the conversion of one kind of energy into another, are susceptible of explanation by the universal principle of allotropic change—not from one imponderable agent into another, but of a change into five distinct modes of manifestations.

The dividing line between the animate and the inanimate, the living and the dead, the line between the organic and inorganic, is definite, abrupt and absolute, wholly and entirely dependent and founded on the principles of allotropic conditions.

Heat that is radiated on the untamed energy of a sunbeam, or the rays of a burning fire, becomes in another condition the mild, genial warmth of animal life.

Electricity, which in one condition dashes with lightning glare from cloud to earth, rending and tearing both organic and inorganic, becomes in another condition the harnessed steed that a child may guide through all the intricate turns of a metal wire.

Light, which in one condition dashes through space two hundred thousand miles a second, becomes in organic life the enrolling agent of instinct, sensation, thought and memory, as well as the induction spark for allotropic transitions to determine locomotion and growth.

Around this central fact of allotropic change, and conditions are grouped the molecules and forces of organic life.

The general reader might at first form a more definite conception of nerve force if I had confined this discussion of the force entirely to phenomena that are the direct results of nerve influence; but as the force that traverses the nerves of animal life is a force that is no more confined to the nerves than electricity is confined to telegraph wires, no more confined to nervous matter than heat is confined furnaces, a discussion of nerves would give but a very narrow conception of the wonderful adaptation of sunlight to organic life, or the broad range of the agent contained in nerves.

Those who are at all familiar with the structure and functions of animal life, know that the human system, and animals that have a back bone, are provided with an elaborate system of nerves leading from the spinal marrow to every part of their bodies—not only to the skin, but the teeth tissues, bones, muscles, arteries, veins, heart, stomach, every part is provided with pairs of these lines through which both sensation and force are transmitted.

The function of digestion is controlled by the force that traverses the nerves of the stomach: breathing, the beating of the heart, motion of the arm, and all vital as well as muscular motions are so completely linked with and dependent on the agent that traverses the different nerves that the study of any phenomena of organic life necessitates a study of the influence and relations of nerve force.

Physiologists have discovered that all the nerves that ramify through the animal structure invariably lead to the different part in pairs: one line of the pair being white, and the other a greyish white fibre. They have also ascertained that the white nerve is more especially the nerve of sensation, and the grey one the nerve line for transmitting force. It is also known that the presence and healthy condition of each of the pair is essential to the performance of a function or vital act.

The force that traverses the nerves of the living animal, directing growth and locomotion, beating of the heart, breathing, digestion, and all vital acts, vanishes from our grasp at death, and we ask ourselves what has become of it?

There are two theories offered at the present day in answer to this question; one theory asserts that the force or influence that traverses the nerves of animal life has no distinct or absolute existence. That all our conceptions of what are called imponderable agents, such as heat and light, and nerve force as distinct existences, are myths of the imagination.

They further assert that these myths have neither form, position, nor tangible existence. That heat and light and nerve phenomena are merely modes of motion, merely relative changes of position of molecules produced by the clash of atoms, that these mythical agents that appear one moment as heat, or light, are modifications of momentum, one modification appearing as heat, another as vitality, another as light, each modification of energy being transmutable into magnetism, heat, nerve influence, or other known agent or throb. That at the death of an animal its nerve force is blotted out and transmuted into some other mode of energy.

In an article on vitality the celebrated John Tyndall says:

"The origin, growth and energies of living things are subjects which have always engaged the attention of thinking men. To account for them it was usual to assume a special agent, free to a great extent from the limitations observed among the powers of inorganic nature. This agent was called the vital force; and under its influence plants and animals were supposed to collect their materials and to assume determinate forms. Within the last few years, however, our ideas of vital processes have undergone profound modifications."

"A few years ago when the sun was affirmed to be the source of life, nine out of ten of those who are alarmed by the form which this assertion has latterly assumed would have assented in a general way to its correctness. Their assent, however, was more poetic than scientific, and they were by no means prepared to see a rigid mechanical signification attached to their words." "To most minds, however, the energy of light and heat presents itself as a thing totally distinct from ordinary mechanical energy. But either of them can be derived from the other. Wood can be raised by friction to the temperature of ignition: while by properly striking a piece of iron a skillful blacksmith can cause it to glow. Thus by the rude agency of his hammer he generates light and heat. This action, if carried far enough, would produce the light and heat of the sun." "If, then, solar light and heat can be produced by the impact of dead matter, and if from the light and heat thus produced, we can derive the energies which we have been accustomed to call *vital*, it indubitably follows that vital energy may have a proximately mechanical origin." Tyndall in Fragments of Science ; article Vitality.

The other view of light and heat and all imponderable agents, utterly discards the notion that a fraction of heat, or light, or vitality, or any other so-called imponderable agent, ever has been, or ever can be generated by the blow of a hammer, the combustion of elements, the impact of planets, condensation of nebula, or any other possible clash of atoms or worlds. It asserts the absolute existence of these agents as inconvertible, indestructible existences, as mutual counterparts of material substances, and declares that the movement and phenomena both of organic and inorganic matter is governed and controlled through the medium of these agents. This view of imponderable agents asserts that the so-called generation or production of heat by a hammer's blow, is only a mode, and only one of the several modes, of developing heat that already exits; and it further asserts that no vitality or other imponderable agent can be produced or generated by the clash of atoms, in any other sense than to render manifest what already exists ; that all of these agents are already in existence and may be developed and guided into new channels to influence and control in succession different forms of matter.

In this discussion of nerve force this last view is the

one adopted, and nerve force treated as an absolute indestructible existence.

The analysis of the phenomena of life, shows that what was formerly called vital force, instead of being a single and distinct force, is an aggregate of several distinct forces, each having its own peculiar modes of development, and producing its own peculiar phenomena.

That heat as a distinct force is essential to animal life is generally admitted, and can be easily proved; but that animal heat is different from sun's heat, or the heat of combustion, has never been asserted or suggested; and so of the force that traverses the nerves of animal life, what is here asserted of this force is that it is simply an allotropic condition of light; that the agent is identically light, in the same sense that heat contained in water is identically the same agent that will cause expansion of a bar of iron.

The discussion of the transmutation of forces is omitted here, for the reason that there is not an unchallenged fact yet offered in support of transmutation.

The assumption that vital actions of organized life are brought about by the drifting impact of material masses, atoms, or ethers, impinging on complicated combinations of matter to produce the functions and phenomena of human life is too indefinite for discussion.

Light from the sun impinges on the sides of mountains, on the earth's surface, on the waters of the ocean, and at first view, seems in many cases to be blotted out of existence. We may light, and put out an electric light, or even a common gas light hundreds of times a minute and each time that it is re-lighted, it sends rays of light in every direction, filling the entire unobstructed space for miles, and we ask ourselves, what has become of the light thus distributed by the ignited gas or incandescent carbon? or what becomes of the impinging sunlight that is constantly received by the earth? When sunlight impinges on a lake or body of clear water the fluid becomes illuminated to a considerable depth, it is not all blotted out at the surface as appears when it strikes the solid earth, but fades out gradually. A bright coin, fish and other objects in clear water may sometimes be seen twenty, thirty, or fifty feet below the surface; the greater part of the light, however is absorbed near the surface and as it penetrates the fluid it all gradually becomes absorbed, but the moment that the source of illumination is shut off, the water becomes blackened with darkness, just as a room becomes blackened with darkness, so that the question still arises, what has become of the absorbed light?

There are some bodies, like the diamond, loaf sugar, some kinds of spar, and others, that will absorb light, and then when the source of light is cut off, will emit the absorbed light for hours afterwards, but the earth, water and most substances will continue for ages to absorb and still not emit the absorbed light, and it is of these we ask ourselves what has become of the absorbed light?

From what has already been said of the allotropic conditions of force, the reader will not be surprised at the position taken, or the views now offered respecting the mysterious energy, or agent, light.

This position is, that the force that traverses the nerves of animals is an allotropic condition of light.

Allotropic Conditions of Matter and Allotropic Conditions of Force.

In order to follow or trace the nature of light in nerve force with satisfaction, the general reader should understand what is meant by allotropic conditions of matter and allotropic conditions of force, and it is desirable that there should be no hidden or mystified meaning attached to the term allotropic.

As applied to material substance, it simply means that the same substance may and does exist in different conditions by which they manifest different properties. Carbon, for example, may exist in the form of glittering diamonds, or it may exist in the condition of coal. Sulphur is familiar in three conditions—flowers of sulphur, stick or roll sulphur, and as flexible sulphur wax. Phosphorus is known as stick or common phosphorus, red phosphorus, and in another state as fluid phosphorus.

Oxygen is known as atmospheric oxygen, and as ozone. Water is known in the fluid condition, in the solid condition, as ice, and in the gaseous condition, as steam.

These are well known conditions of the several bodies, that are convertible conditions, as of ice into water, and water into steam, and steam into ice or water.

Allotropic conditions, as applied to forces, or imponderable agents, means simply their different modes of manifestation. Thus heat is manifest in the sunbeam, traveling two hundred thousand miles a second, the radiant mode. It is manifest when passing along a metal rod less than a foot a second, the conducting mode.

Heat is also known in the dormant, inactive condition, as in ice water.

Electricity is familiar as surface electricity, and as current electricity.

Magnetism is familiar as localized in steel magnets, and in another condition, as enveloping the earth, and in another condition as conducted by, or with voltaic currents. Light is well known as fluorescent, phosphorescent, and radiant light.

This brief statement gives but a mere glimpse of the allotropic conditions of matter, and force. The primary elements that enter into and form the structures of organic life have five distinct states or conditions, and the forces that are manifest in the processes of organic life have five distinct modes of manifestation.

Material substance, and immaterial forces are mutual counterparts. Whenever a primary element absorbs and retains an imponderable agent, the agent produces an allotropic change in the condition of the substance; and whenever a material substance discharges an immaterial agent that it has absorbed and retained, the allotropic condition of the imponderable agents becomes changed and it becomes manifest in a different mode of manifestation.

This brief summary of allotropic conditions, and allotropic relation of forces and matter, will prepare the reader for the statement now made, that *Light*, as well as all other imponderable agents have five distinct modes of manifestation.

What is here asserted of light in this connection is that light passes from the radiant mode as manifest in the sunbeam, and becomes manifest in each of the other modes in organic life; that it traverses the nerves of animal life by conduction by virtue of its polarization, but that the same force may be traced in the vegetable in their molecules, and in their instincts; that the nerve current embraces each and all of the members of the sunbeams group—its heat, its re-arranging force, its obscure and nonluminous rays, and also its property of enrolling knowledge.

From this brief statement of some of the allotropic states in which the primary elements are known to exist, it will be seen that the same elements not only present differences of molecular structure, appreciable by the senses, as of hardness, color, flexibility, crystalline structure, &c., &c.; but both the elements and their combinations exhibit different affinities, being in one allotropic condition devoid of chemical affinity, while in another condition they show an intense affinity. The allotropic conditions of matter and the allotropic conditions of force forms the dividing line between the organic and inorganic world.

The change of matter from one allotropic condition to another develops the forces engaged in the various processes of organic life. The allotropic condition of matter of organic substance being totally different from the allotropic condition of inorganic substance.

Allotropic Properties of Forces.

At first view imponderable agents in their different allotropic states seem like distinct agents, but as we examine them attentively we discover analogous or homologous properties, and it is a well recognized fact that each imponderable agent is adapted to a particular class of duties, and each agent performs its own peculiar work; so that magnetism adheres to its own duties, and never expands bodies like heat, and heat never attracts bodies like magnetism. If we turn to the phenomena of light and fix our attention on some prominent property or properties of light, and assume that this prominent property exists in each of the allotropic conditions of light, then the property in each phase would have a marked family resemblance.

If we ask ourselves, what is the most prominent property of light, the answer that immediately occurs is, the property of illumination; the property of rendering objects visible; forming a medium of communion between matter outside of ourselves and our perceptions; a medium that enables us to apprehend the presence of distant objects. And if we turn to our nerve force, we find that it also enables us to apprehend and recognize the presence of objects at the different parts of the body; a percipient medium between our inner selves and what will cause pain or pleasure on our outer selves.

In other words, light outside of the body is a medium of intelligence, and an allotropic condition of the same agent transmits intelligence within the body. Light gathers, enrolls, and transmits information to the body and the nerve force performs identical duties within the body.

Light Gathers Information, Enrolls Knowledge and Embodies Actual Force.

There are three common and well known properties of light that are well known properties of nerve force, and can also be traced in other allotropic conditions of the agent.

These three properties are, property of gathering information, enrolling knowledge, and manifesting actual force. And these are the prominent properties of nerve force.

In analyzing light, we find that these three qualities, or properties, may be separated and detached from each other. And we also find that in the nerve force these qualities are separated by polarization; the sentient side or quality being found in the sentient nerves, while the qualities of force are found in the motor nerves, the two classes of nerves having mutual poralized relations to each other.

Light itself is invisible, no more to be seen than heat, or magnetism, or even the ultimate essence of life.

It must have something to shine upon to become visible, and in this case we do not see the light, but we see the object illuminated. It is the agent that reveals their properties of form and color; it carries information and knowledge of their qualities. A beam of sunlight is unseen in empty space, but after it enters the atmosphere it gradually becomes luminous or illuminating, and in acquiring this illuminating property it becomes polarized.

Polarized light is light split and divided into two conditions, each part or condition having distinct qualities.

Previous to its polarization we know as little of its ultimate essence as we do of unpolarized electricity. We know absolutely nothing; but in the polarized state a sunbeam is the embodiment of most, if not all organic agents; but we know of them only in their polarized state by their influence and actions, and not in their previous state. The influence and manifestations of not only light but of all imponderable agents, is founded on their polarized conditions, and in and around this central fact all known phenomena of electricity, of magnetism, of heat, instinct, intelligence and life are grouped.

By the polarization of light two prominent facts appear: one is that one side or condition of the polarized light reveals to sensitive matter, the properties and qualities of objects at a distance; the other fact is, that the other side or condition of the polarized ray induces action in matter in accord with the sensitive prehensions or informations respecting qualities and properties of distant objects. One side of the ray carries information and knowledge of qualities or distant matter, and the other side carries force suitable and adapted to render that information and knowledge useful.

These two conditions of a polarized ray are as readily separated in position from each other as the two polarized conditions of voltaic electricity are from each other; and, like the two conditions of electricity they cannot be entirely detached, but will maintain a correspondence and relation to each other; the positive or negative condition of one battery refuses to work with a counterpart condition of a separate or detached battery, it will only perform work in conjunction with its own counterpart. And so of a ray of light, the two conditions cannot be so separated as to destroy their correspondence and relation to each other.

Light Embodies Both Knowledge and Force.

The gases and motes of the atmosphere polarize light and impart to it its illuminating quality and give to the sky its blue color. The atoms of the atmospheric gases that produce this effect are too small to be visible, but their united effect produces day light. The entire mass, however, of sunlight does not become polarized by the atmosphere, but only that portion that renders objects visible, while one side or part of a ray of polarized light is carrying information of qualities, and the other side is carrying a condition of force they may be moving in quite different directions, substantially as wires from a battery may carry the current or the two conditions in different directions, but in performing work they must meet and unite again somewhere, and by their union their manifestations occur. Each of these polarized conditions of light may become latent and inactive; the sentient side as knowledge in memory, the force side or condition in vitality, as exhibited in dormant seeds and eggs.

In its allotropic state as nerve force, the two polarized conditions of light are to be found in the two classes of nerves, the sentient and the motor; substantially as the two conditions of electricity are separated, the positive, on one plate or wire, and the negative on the other plate or wire. Light becomes polarized by refraction, by reflection, and by diffraction, and it would seem that the atmosphere polarizes it by each of these modes, and that each of these contribute a shape in giving to light its properties of being the medium of knowledge and force.

If we place a penny or other piece of copper in a burning fire, and watch it while it is being heated, we shall see it emit a bluish green light just before it becomes red-hot; very slight portions of the copper are carried up with the coal flame, and it is this burning copper that colors the coal light; to the eye of the experienced metallurgist or copper worker this greenish flame is positive evidence of copper in the fire. There are no particles of copper transmitted to the eye to produce this result, but when the colored light is seen by the eye of the worker he perceives the fact of burning copper, and this knowledge and information is transmitted to him by light. If, instead of copper in the fire there had been lead, or iron, or sodium, or some other metal, the light emitted by their burning would have shown a distinct color for each, and would have carried to the eye correct information of their burning. The light emitted by the burning copper not only reveals to the eye what is burning, but the light also embodies heat and chemical force in its rays.

Light Enrolls Knowledge and Force.

The illuminating agent brings information to the eye that is absolutely true and perfectly reliable. This information is transmitted not only a few feet, rods, or miles, but the information may come millions of miles without loss. Rays of light emitted through burning copper or other metal or element will bring the information millions of miles. A bit of burning copper no larger than a pin's head in the sun will write the information on a ray of light and the knowledge will come with the ray of light to the earth.

When light carrying information of this kind is passed through a three cornered piece of glass or prism, different parts of the light are refracted more than others producing what is called the spectrum in which the decomposed light is refracted into a series of colors. These colors are crossed by what are called Fraunhoffer lines. These lines have been minutely studied by Bunsen and Kirchoff, and although there are thousands of these lines across decomposed or refracted light, each line was found by them to indicate a chemical change of an elementary atom; and every material element in the process of chemical change when traversed by a ray of light makes a mark or marks across the spectrum of the ray and always in the same relative position—no other element ever making marks in the position pre-empted by another element—each has an appointed place to give accurate information of what it has done and is doing millions of miles distant; and although the courier dashes through space nearly two hundred thousand miles a second, the writing or marks of one element never gets intertangled with the marks of another substance.

Light may come from burning coal, or iron, or earth, copper, gold, salt, any known substance. Each and all will have its own appointed position on the courier beam and its message comes to the earth accurate and perfectly defined. Although there are thousands of these Fraunhoffer lines across a beam of light, the beam when spread out by refraction reveals each line in its true position. These lines are sometimes bright, sometimes dark or with intermediate colors and shadings, characters which have been shown to be produced, and indicate the character of chemical change, whether of intense combustion and emitting their own light of combustion, or whether they are absorbing light and information from some other source of light—bright lines indicating their own combustion. These Fraunhoffer lines and spaces between give infallible information of the chemical changes and character of substance; they tell us of what elements they are composed, or through what substance light is transmitted, whether it is sodium, potash, lead, iron, carbon, hydrogen, &c., but this is a kind of information that is more especially important to the chemist—it is information of the constitution and internal character of substance and is information of the inorganic world. But light not only brings information of the internal constitution of matter, but it also brings information of superficial and external qualities of subjects; it not only tells of what the earth is composed, but it also tells whether it is covered with snow, or with trees, of green fields, of houses, of animals or people; it brings information of all the intricate qualities of forms, faces, colors, and manners; everything seen by the eye are accurate representations of distant objects; and these delineations are wholly drawn by one side or condition of a polarized sunbeam. The delineations by polarized light consists of information of external qualities, while the Fraunhoffer lines are descriptions of internal character, and the external descriptions of qualities transmitted by the courier are not less accurate or reliable than the descriptions of internal constitution. When the eye is directed towards an object, an exact representation or image of the object in miniature is formed on the retina, and this image is drawn by one side of a series of divided rays. This property of one side of rays of light, describes and tells of qualities of horses, trees, people, or other objects at a distance; it embodies knowledge without force, information without material change. Its relation to material change is somewhat like our conceptions of induction. When we bring a magnet near a piece of iron, the iron will seem almost conscious of its presence—the soft iron may not be sensibly moved from its position, but the magnet has induced a polarization of its substance. A cold body by its action would seem to be half conscious of the presence of a near hot one, but in these phenomena of inorganic actions, it is the force side of a polarized agent that induces action. The eye is but one of the windows or organs of sense through which knowledge is received. Each of the other senses receive information of distant objects through the medium of polarized light, but the mode in which information reaches them is founded on the allotropic conditions of light. It has other modes of traveling besides the radiating. When a mirror is held before objects, we see images of them reflected from its metal surface. The radiating images of the mirror carry no information that can be recognized by the end of the tongue, finger tips, or nerves of the nose.

The ear is entirely unconscious of flowers, mountains, or rainbows. The allotropic mode by which knowledge is transmitted to the eye, is incompetent to give information to the nerves of the ear or either of the other senses; yet knowledge imparted to the ear by sound, to the hands by touch, to the tongue by taste, or to the nose by scent, is as accurate as that through the organs of sight.

Each of the several organs of sense is adapted to receive knowledge and information from surrounding objects by one, and only one, of the allotropic states or conditions of nerve force. As already noticed, imponderable agents act by radiation, undulations, conduction, convection, and by disintegration; and in the phenomena of human life the senses are adapted for receiving impressions by each of these modes.

To the eye knowledge comes on the radiations of light—it is adapted to the radiating mode; to the ear, knowledge is imparted by the undulatory mode; the thunder's reverberating roll, the tiger's growl, the harmonies of music, and the gentle wooings of love reach the drum of the ear on the undulations of material substance. And here comes a suggestion that may at first appear erroneous and unnecessary. It is that knowledge derived from sounds is loaded onto, and carried by vibrations, but that which reaches the nerves is more subtile than vibrations, precisely as information that reaches the eye is not mere radiations or undulations of ether. The subtile agent that is carried to the ear by atmospheric vibrations brings harmonies, melodies and discords in all their intricate modifications that represent properties, qualities and capacities of organic life, which are received and folded on the leaves of memory, remaining dormant and inactive as a printed page of music, to be recalled like the pictured forms and faces of vision, at the option and command of consciousness.

These subtile qualities and properties of distant objects, even while being carried by vibrating waves of sound, may be captured by an electrical current, detached from the vibrating substance and carried on telephone wires, or through the earth circuit without an appreciable vibratory swing, hundreds of miles. The vibratory pulsations of sound through material substance are the manifestations of the force side or condition of a polar force, and this force or vibrating condition is linked and accompanied with the sentient side or condition of the agent.

The discussion of the idea of the reception and retention in dormant inaction by either of the senses of hearing or seeing, of images, and qualities of distant objects, opens up the discussion of the principle that underlies the undulatory theory in all its ramifications. Those that can perceive in heat only a mere mode of motion of material substance, a mere shimmering motion derived from the clash of atoms, and ignore the sensations of warmth imparted to organic life as intangible myths, or can trace from light nothing but undulations of an attenuated ether, ignoring those latent images impressed on the canvas of memory through the medium of the eye—images that are subject to the roll-call of consciousness—will see in the vibratory motions of sound, only the pulsatory movement of material molecules, and will look on the latent images of vision, or the treasured tones of the dead when recalled from the realms of dormant sleep, as mere mythical echoes, or non-existent delusions. Yet we are surrounded by analogous phenomena of the junction of the imponderable with the ponderable, not only in organic, but also in inorganic matter.

When a magnet is moved near a piece of iron, copper, or other metal, the minutest motion of the metal induces invisible movements of imponderable agents conjointly with the ponderable. Electricity, and heat, and light, in movements far more subtile than the most delicate vibrations of the metals are conjoined with these movements, and are as absolute as the perceptible ones; or when water or other substance is evaporated, the volatile molecules are conjoined with imponderable heat, and both the ponderable molecules and the imponderable heat reach the sense of feeling, but it is only the imponderable that imparts the sensation of warmth.

The interchanging material elements of a galvanic battery are conjoined with the transfer of imponderable agents more subtile than the material atoms interchanged; and so with sound, that which reaches consciousness has been carried by the vibratory condition of matter, but the information imparted through this sense is more subtile than mere vibrations. Each distinct sense has distinct modes of receiving information and knowledge. The sense of smell is wonderfully accurate in dogs and some of the lower animals. Information and knowledge is conveyed to this sense by convection. The qualities or properties of objects are loaded on volatile molecules and brought in contact with the nerves of this sense. Then again information and knowledge reaches the sense of taste by still another mode; taste depends on solution and disintergration of substance holding properties imparted by the sun's rays in a dormant state of inaction.

If there is anything in the phenomena of animal life that is calculated to excite wonder and admiration, it is the fact that impalpable daylight should be concentrated into a condition to traverse the nerves as nerve force. Yet evidence of this fact gathers and accumulates from a great many sources.

We can trace in an almost unbroken line the influence of light on the inorganic matter of the earth; the influence of invisible rays of heat from the sun on the roots of plants, and the influence of visible rays on the leaves; the absorption of both visible and invisible in the organism; the direct influence of light on leaf and flower, and the direct influence of light on the organs of vision, and on the skin of animals.

The direct influence of light on the skin or eyes of

animals frequently produces effects so closely resembling those produced by the nerves that different observers differ as to whether they are caused by direct light, or by the nerves. This is the case in the movements of different parts of the eye; the humors, coatings and lens of the eye have the property of spontaneous adaptation to near or distant objects, and to bright or dim lights.

A person standing before a looking glass in a dimly lighted room can see the curtain or colored part of his eye spontaneously shutting off a part of the light when the light of the room is suddenly increased.

The remarkable effect of light on those small lazards known as chameleons, has in all ages excited the wonder and curiosity of the student. Their skins acquire a great variety of rainbow tints, both by the influence of light and also when irritated. Their colors rapidly change from grey to yellow, green, brown, violet, with intermediate spots and changeable variegations. If one part of their bodies only is exposed to light, that part, or perhaps some other part, becomes spotted. If one eye is exposed to light some part of its body changes its hue. Somewhat similar changes of color, in an inferior degree may be seen on the skins of the common tree frog, as well as many water animals. Light, the electric spark, a live coal and other irritants induce changes in various colors in these low animals, and it might be a difficult question to decide whether the effect is from direct light, or from the influence of nerves, or whether both direct light and indirect light through the nerves contribute a share; but in either case the effects are at the border line between direct radiations and the allotropic condition of the same agent by conduction through the nerves.

The influence of light, in changing the colors of flowers and leaves of plants, and its influence in changing the color of the skins of animal life is not phenomenal or confined to low orders of animals, but affects the highest, while the sudden change of color from rage, hate, grief, joy, anticipation, or disappointment through the nerves is familiar to all.

In order to form a true conception of what light is capable of doing in its allotropic condition of nerve force, we must fix in our minds the fact that it is an indestructible energy, constantly existing in a variety of forms; sometimes radiating through space with inconceivable velocity; sometimes permeating the soil at an almost snail like pace in conjunction with warmth; sometimes moving through the pores of vegetation, or the veins of animal life in conjunction with their vital fluids; sometimes traversing the pearly nerve fibres as quick as thought, and reaching remote parts of the system along these conducting lines, and then also remember that it is sometimes as inactive as the latent heat contained in a pail of water. These several modes in which light becomes manifest, carry the three qualities of gathering information, enrolling knowledge, and embodying force, and enables the self-same agent to produce phenomena of intelligence and power in organic molecules and structures devoid of nerves—vegetable instincts and forces being founded on allotropic conditions of the same agent that imparts animal intelligence

The allotropic conditions of light are manifested in five distinct modes, and it is a significant fact that these five modes correspond in number, and are adapted to the five senses of man. These five modes are the radiating, the permeating or vibrating, the convecting, the conducting and the latent. The radiating mode is adapted to vision, to the sense of seeing; the permeating or vibrating is adapted to the sense of hearing; the mode by convection is adapted to the sense of smelling; the mode by conduction is adapted to the sense of feeling, while its existence in the latent mode is adapted to the sense of taste. In considering these five modes, we must bear in mind that it is only the illuminating part of the sun's rays that render objects visible to the eye; and that less than one-third of the sunbeam is adapted to the sense of seeing; and we must also bear in mind that the sun's rays in space are non-luminous; that they must impinge on the substance of the atmosphere to acquire the illuminating quality adapted to the sense of vision. By keeping these facts in view, we shall be better prepared for the statement now made, that the invisible rays acquire an adaptation to each of the other senses by transmission through other conditions of matter: the transparent gases of the atmosphere adapts a portion of the sun's rays to vision, but that other forms and conditions of matter adapt the non-luminous rays for transmitting the

What is Light?

The question is frequently asked, what is light? And many attempts have been made to answer it. By some, light is called an attenuated substance; by others it is called an attenuated motion. The difficulties connected with either view are appalling; and it is doubtful if either the emission theory of an attenuated substance, or the undulatory theory of an attenuated motion gives us any deeper or clearer insight into the nature of light than is obtained by simple observation. Our eyes, by the emission theory, are pelted by the attenuated atoms of matter discharged by combustion and impinging on our optic nerves with a velocity of one hundred and eighty thousand miles a second. The battering force of these impinging atoms has been variously estimated and calculated by mathematicians, and by these estimates the impinging force of the atoms would far exceed the force imparted to matter by gunpowder. The most attenuated matter known is hydrogen, and if this substance was driven with the velocity of light against our bodies, they would melt, vaporize and vanish in thin air in a very few minutes.

We cast our eyes around us in every direction, and light from millions of objects reach our vision, depicting their forms and colors on the retina. The light for these images may pass through glass, water or other transparent substance; it may be sifted, or it may be absorbed for days or years, and yet no substance that can be called light has ever yet been caught for examination. Then again substance may be burned in a glass vessel emitting an intense light in every direction, and the carbonic acid or other material products of combustion examined, and the original amount of carbon and oxygen, or other substance burned is all there in the vessel, not the minutest portion having escaped with the emitted light.

From these and a variety of other considerations a great many scientists have discarded the emission theory of light and adopted the undulatory theory under the delusive notion that matter and motion comprehends the entire circle of all existence.

According to the undulatory theory space is filled with an etherial substance, and vibrations of this substance constitute light. Teachers of this theory present definite statements of the motion which the hypothetical ether must make to enable the eye to see.

The following is a table of vibrations, claimed to be necessary to form the colors of a bouquet of a few flowers, to render them visible. The table is from Guillemin's School Philosophy, and is the doctrine commonly taught in schools.

Vibrations Per Second.

Red, 514,000,000,000,000.
Orange, 557,000,000,000,000.
Yellow, 548,000,000,000,000.
Green, 621,000,000,000,000.
Blue, 670,000,000,000,000.
Indigo, 709,000,000,000,000.
Violet, 752,000,000,000,000.
—[Guillemin's Forces of Nature.]

To understand this table we will suppose that a spider's thread is drawn out and attenuated, in imagination, to the least possible tenuity and reaches to the sun, or to a burning gas, and that a shimmering or vibrating motion is imparted to the thread by combustion. This vibratory shiver of every part of the thread, which is 93,000,000 of miles in length, each inch must vibrate at least five hundred and fourteen billions times a second. Any less number of vibrations or undulations will not rouse the sense of vision. Every forty-thousandth part of a red rose blossom is assumed, by this theory, to be sending off these billions of vibrations in every and all directions. In order that light shall travel over a space one inch long and the one forty-thousandth of an inch broad, the substance of the ether must make forty thousand oscillations. The whitish yellow light of a lamp, as will be seen from the table, vibrates still faster than the light from a rose blossom, while to render a violet colored blossom visible, there must be 752,000,000,000,000 vibrations in every second, or about sixty thousand in every linear inch to illuminate one sixty thousandth of an inch in breadth.

Professor Tyndall, one of the leading exponents of the undulatory theory, says, "The length of the waves, both of sound and of light, and the number of shocks which they impart to the ear and eye have been strictly determined."

"Let us here go through a simple calculation; light travels through space at the rate of 192,000 miles a second; reduce this to inches, we find the number to be 12,165,120,-000. Now it is found that thirty-nine thousand waves of red light placed end to end, would make up an inch. Multiplying the number of inches in 192,000 miles by thirty-nine thousand, we obtain the number of waves of red light embraced in the distance of 192,000 miles; the number is 474,439,680,000,000. All these waves enter the eye in *a single second*."

"To produce the impression of violet, a still greater number of impressions is necessary; it would take 57,500 waves of the violet to fill an inch, and the number of shocks to produce this impression of this color amounts to six hundred and ninety-nine millions of millions per second. The other colors of the spectrum, as already stated, rise gradually in pitch from red to violet."—[Tyndall in Heat as a Mode of Motion.

Again, " you must then imagine the atoms of luminous bodies vibrating, and their vibrations you must figure as communicated to the ether in which they swing, being propagated through it in waves; these waves enter the pupil across the ball and impinge on the retina at the base of the eye, and the act, remember, is as real, and as truly mechanical as the stroke of sea waves on the shore; the motion of the ether is communicated to the retina, transmitted thence along the optic nerve of the brain, and there announces itself to consciousness as light."—[Tyndall in Heat as a Mode of Motion.

These estimates, it will be noticed, are only for the path of a single ray of light; but each ray illuminates a streak or path only the thirty-nine thousandth of an inch in diameter; and as the pupil of the eye usually presents an opening about the tenth of an inch in diameter, it enables several thousand of these rays to enter side by side in their inconceivable oscillations.

We seat ourselves in a room that is perfectly dark, and we see nothing; the ether is all there permeating the entire space, but it renders nothing visible. We open the window shutter, and let in daylight, and then, by the theory, every minute part of the floor, of the walls, ceiling and air of the room, sets the permeating ether quivering. We hold a small bouquet of flowers, having different shades of red, orange, yellow, green, blue, indigo, violet, intermediate shades of bright, dark or faded parts; each of these intermediate shades of which several hundred may be counted on a small bouquet, is sending off billions of vibrations, each different shade of color sending a different rate from either of its mates; but each shade sets the ether vibrating more than 514,000,000,000,000 times a second, ranging from that up to the violet with 752,000,000,000,000 undulations a second, and a few only of these uncounted billions are dancing on the optic nerve to enable us to see a few flowers.

It might be as well to say here to students, before loading their brains with this avalanche of figures in explanation of light, that the ether that vibrates so rapidly has not yet been discovered; and that the vibrations of different colored lights are only estimated ones, founded on the proposition that if the *ether* does exist, filling all space, and if light does travel through the ether in this way by undulations, then such must be their mathematical ratios.

This theory of light asserts that black is not a color; that black objects absorb all kinds of rays, but emit none; that black bodies will not set the hypothetical ether vibrating. A very natural conclusion from this would be that black bodies should be invisible; but when we stand before a looking-glass we see images of all kinds of objects; we see black eyes, black hair, and black clothing quite as plainly as we do those of other colors; or if from an upper window we cast a glance along a busy street in a populous city, we receive a momentary impression of its long rows of buildings, stores with their windows full of the useful and beautiful, streets full of a busy throng; horses, carriages, men, women and children—forms, figures, faces in infinite variety, with shadings of all colors blended and intermixed with the vital throb of active life.

In the phenomena of vision, images in miniature are formed on the inner lining of the eye, faithfully representing in the minutest detail each and every form, face, figure, of these and other objects with their distinct or blended shades of coloring. The particles of soot that lazily curl from the chimney tops, although they are supposed not to stir the etherial substance with a single undulatory shiver, appear on the canvas of the retina defined as accurately as the snow white lime that cements the chimney joints.

The undulatory theory of light, as well as other attempted explanations of the nature of light, are efforts of the human mind to account for all classes of images of distant objects that are reflected from mirrors, refracted from lenses, or formed on the optic nerve, and impart knowledge of their qualities.

After a momentary glance at the unnumbered objects of a busy street, at the whitened sails in its harbor, at the waving foliage of a forest, or the vivid bloom of flowers, we retire to a darkened room where the hypothetical ether is supposed to be at rest, and the images which light penciled on the retina, have vanished, the canvas is ready for another series of pictures. The undulating pencils have ceased to dance on the nerves of the canvas; we close our eyes in forgetfulness, and by the etherial balm of sleep become oblivious of either undulations or of objects. After a period of repose we awake and recall from the tablets of memory, the whitened sails in the habor, the waving green of the forest, the brilliant bloom of the flowers, and the active throbbing life impressed in the momentary glance along the throbbing street.

In this process of recalling impressions of objects and their qualities from memory, the images do not again appear on the canvas of the retina; neither does the undulating ether return to dance on the optic nerve; the intermediate light between the eye and those distant objects from which the pictures were drawn has vanished, and it is unnecessary for their reproduction from memory. The houses, people, sails, forests and flowers may have vanished in smoke; their presence is also unnecessary in the recalling process, and we ask ourselves what is the real difference between the pictures from memory and those first formed by the illuminating agent?

If there was any conceivable relation between knowledge and vibrations, any relation existing between consciousness of the existence of any outside object, and the undulations of an intervening medium between these objects and the eye, any positive dependence of vision on motion, the theory of etherial undulations would be admissable whether it could be discovered as a fact or not; but no such relation is either known or conjectured; vibrations, shimmer, quiver, thrill, undulations, or whatever it may be called, of either the substance of ether or the imparted motions to atoms or nerves, is entirely inadequate to constitute vision. The motion and vibratory shimmer which the substance of ether has been computed to make is inconceivable, undiscovered, and seemingly beyond the power of discovery; and should it be discovered the contrast between perception and the undulations of substance is abrupt and disconnected. Motion, however rapid, does not become knowledge; the beating of the waves on the sands of the sea shore does not impart knowledge or sensation to the crystalized silica; the twinkling of the stars, or the undulations of a sunbeam on the ocean, or on the deadened nerves of extinct life will not confer vision on the trembling molecules.

Discarding all theoretical notions of what constitutes light, and discarding all theoretical notions of what constitutes consciousness, for the present, and reverting only to primary facts, we know that the earth exists; that its forests, its people, its cities and everything around us does positively exist, and we also know that light has somehow brought us a part of this knowledge which we have of these objects.

It is not only true that objects around us exist, but it is also true that a knowledge of these facts also exist, and it is also true that we are conscious that we have this knowledge. During sleep this knowledge lies dormant, but when we awake we are again conscious of these existences. This conscious knowledge is as real an existence as is the earth on which we stand. At the demand of consciousness, we can recall knowledge derived through the eye, and in discussing analogies, or differences existing between the qualities of objects which we see and qualities of objects which we can recall from memory, the position is presented that the process is accomplished in both cases by the same agent; that light draws the pictures on the canvas of consciousness, and that light recalls them to view; that the same agent that enables us to see objects and their qualities, is the agent that enables us to recall them from memory. As already presented, this is accomplished by the allotropic conditions of the agent.

The property of light to render objects visible is undoubted, unquestioned and absolute. Theories may differ as to how this is accomplished, but the fact is universally conceded; the power of recalling knowledge from memory possessed by the human mind is also undisputed, unquestioned. and absolute, and this power is generally conceded to be dependent on the nerve force. In the first case consciousness is reached by light traveling through a medium outside of ourselves; in the second case consciousness is reached by light in its altered shape, reaching consciousness by traveling through the nerves as a medium. The accuracy of this statement is found in the fact that luminous rays are perfectly competent and have the property of carrying absolute information of forms and colors of distant objects through the air, water and other transparent bodies; and in the corresponding fact that the nerve force, in what is here claimed to be an allotropic condition of light, is competent, and has the property of holding and carrying the same absolute information of forms and colors of objects along the nerves.

Vision.

We stand by the side of a person in front of a looking-glass and turn our eyes alternately towards the person and to his image reflected from the mirror, and perceive that his form and various qualities which come from direct observation is also accurately reproduced by the reflected image. The accurate information which direct light brought of his form, features, colors of hair, eyes, clothing, &c., is not changed, but duplicated by impinging on and passing through the glass to the metal, then rebounding from the metal and again passing through the glass towards the eye. Both the direct and the reflected images pass in succession through the air, then through the outer coat of the eye, then through the pupil, then through the crystalline lens, through the aqueous and vitreous humors, each successive part contributing an influence towards forming the concentrated miniature images on the retina.

The minature image of a person may be but the fraction of an inch in length, with accurate proportions of every perceptible part, from the minutest hair or expression to the entire outline. A black fly may be resting on a black button, on a black coat on the person, and its miniature is pictured as perfectly accurate as the larger miniature; light has somehow transferred to the retina pictures of these and other objects as an essential fact in the phenomena of vision. The formation of these images on the inner lining of the eye constitutes a part of our positive knowledge. We see objects and we know that they exist in position and character as these images declare them. We turn our eyes towards a tree in the garden, a horse in the street, a fish in water, or the moon in space, images are formed in the eye, and we know that these objects exist in the several places declared by the sight. We also know that light, carrying information of the forms and qualities from these and other objects, may be reversed and turned in various directions, reflected, refracted, and may also be stopped before it reaches the eye and made to explain what kind of images, and what kind of information it is carrying.

If, instead of passing through the crystalline lens of the eye, the luminous messenger with its essence of knowledge from a tree, or other object which it is carrying, is passed through a glass lens, a miniature representation of the information which it is carrying is drawn at the focus of the lens; if the apparatus of the glass lens is as perfectly adapted for diminishing and concentrating information and knowledge, as is the apparatus of the eye, the pictures at the focus of the glass will be as perfect as those formed on the retina.

The information concentrated in the images at the focus of a glass lens is not drawn or engraved by a material substance; its colors are not paints or dyes; its outlines are not inks or leads. The image may be of a tree loaded with fruit, yet no part of the tree, not a molecule from a leaf, not an atom of the fruit or of its aroma is taken or to be found in the image at the focus; neither is the image known to embody any kind of motion from the throbbing life of the living tree; the images are as true from inanimate as from animate matter; it draws on neither molecules nor atoms for the motion of its pencil or brush; and there is no known necessity for asserting the messenger to be either matter or motion—either substance or vibrations. This messenger of knowledge, and this messenger of force, as agents of the infinite, are untrammelled by space, and oblivious of time—their range is the universe, and their life period eternity.

What is known is, that the imponderable messenger of knowledge has the marvelous property of gathering information of the forms, features, and qualities of every known thing in every part of creation, and telegraphing this information all over the universe. A ray of light sent from the sun to the moon goes bounding back to the starting point with the information that its surface is covered over with redish colored rocks of all dimensions thrown around in wild and dreary confusion. When this messenger of the universe, with its load of information encounters a mirror, or reflector in its path, its course is turned, but its information is not confiscated by the mirror; its panoramic pictures are instantly spread out for inspection, reaching apparently into the far distant background of the mirror, but these telegrams of facts, these pictures of objects, these images on the scroll of light, were destined for the sense of vision, and the panoramic scroll, with its marvelous pictures finds an uncrowded depot on which to unload its concentrated essence of knowledge, on the back lining of the sense of sight. A great many scrolls with their panoramic pictures enters the eye as an emblematic gleam of omniscience, but no part of their light is reflected or sent back from the retina; the messenger, after spreading its information on a common mirror, finds no adaptation in its metallic substance; its pictures are not wanted, or retained, and it flies onward, leaving no trace behind. The scroll of knowledge concentrated by glass lenses also continues its flight, carrying its unappreciated knowledge with unchecked speed on its direct or diverted course. Not so however with the panorama that finds entrance to the retina of the sentient sense of sight—the messenger has done its errand of gathering knowledge and bringing it into the living system. The messenger with its canvas of pictures, its panorama of phenomena, has been knocked from its course, now this way, now that, by reflectors; it has been reflected and passed through solid glass, and its scroll opened for inspection, and information, and pictures examined at any part of its route; it has been passed through the membranous tissue of the eyeball, and in succession through solids, liquids and semi-fluids, and in passing through all these different kinds of material, through all its complicated turnings, the messenger clings to its pictures and knowledge, and they adhere to the messenger; they reach the retina together—the panoramic canvas and what it has carried is spread out on the retina, and what warrant have we for separating and detaching the pictures from the scroll? They take up no space, and they require no balm to preserve them; when the messages have been spread out at the focus of the crystalline lens of the eye for inspection, and the facts which they enfold apprehended, we lose direct tracings of the messenger and its load; but the messenger, with its information and panoramic scroll, has proved itself competent to pass through solids and liquids, competent to stand being knocked around in all directions, and still messenger and message, pictures and canvas, enrollment and scroll, phenomena and light, clung and adhered to each other by mutual adaptation and mutual destiny. In the performance of this service light has lost no energy, nor squandered any of its properties; it has been sent from the sun over the intervening distance of millions of miles for the express purpose of making known to sentient beings the nature and qualities of surrounding objects; it gathers information, embodies knowledge, and there is no known separation at the base of the eye ball.

These panoramic pictures, this embodiment of information and knowledge, traced by light on the retina, by the prevalent theory, consists of a quivering motion of an unknown substance; this quivering dance is supposed to be performed by thousands of partners, when an object is seen, their motions determining vision, and each partner in the dance occupies less than the forty-thousandth of an inch of retinal surface, and shakes itself not less than five hundred billions of times a second, while each partner shakes itself several billion times more every second, their vibrations ranging all the way up to over seven hundred billion undulations per second on the nerve. This undiscovered undulatory dance and imaginary thrill goes undulating along the optic nerve to the brain, where it is supposed to be transmuted into knowledge.

But where is the necessity of supposing a transmutation in the brain? What is this fancied transmutation? Do the facts and phenomena presented by a mirror, or unfolded on the retina from a scroll of light, require transmuting into unconjectured pictures? Is knowledge not already enrolled on the scroll of light when it comes from the moon, a tree, or other object? What clearer pictures of objects can the brain form than are already drawn by light? Has nature employed a bungling artist to perform this work, become dissatisfied, and employed a more competent agent to fix them in memory? Are the pictures drawn by light, mere daubs to be discarded and transmuted into something more refined—a thrill to be unravelled by the affinities of brain substance? Are the pencils of light too coarse, the color too crude, or the canvas too destructible to be worth preserving? What transmutations do the pictures of facts and phenomena require to pass into the mind's treasury of knowledge? The canvas is indestructible, the colors are indelible, the outlines are faultless, the pictures are perfect representations of all superficial qualities, and what possible transmutation can they undergo to make them more useful?

When the pictures, enrolled by light, reach the retina, they have the smallest possible dimensions; a landscape view, extending over miles, is accurately reduced on the optic nerve to the fraction of an inch, and so perfect are these miniatures in every particular that a glance at the landscape, and then at the picture within the eye ball reveals the perfection of representation. These pictures of facts and phenomena, have but two dimensions, length and breadth, but they are utterly without thickness.

Millions of these transparencies, embodying all the qualities of knowledge derived through the sense of vision may be hung in the vestibule of memory for reference and comparison and add no more to the bulk of optical substance than is added to a bar of iron or steel when charged with magnetism. They require no transmutation, they are the embodiment of all facts and phenomena that pass through the organ of sight; the scroll of light on which they are enrolled has the property not only of carrying pictures embodying knowledge and information of the qualities of matter, but it has also the property of retaining and holding this information and these pictures for reference and inspection.

When the panoramic pictures of facts, carried by light, reach the retina, the messenger is not dashed into fragments of nothingness, leaving only a few dots and dashes to be transmuted into knowledge by chemical affinities within the brain—the imponderable agent with its load of pictured facts, finds a resting place and home on the sentient nerve for its scroll of knowledge; the pictures do not pass into hyeroglyphics, to be translated by oxidations of brain substance, but scroll and enrollment, pictures and canvas, information and agent, panoramas and light, are filed in the treasury of knowledge and assigned to positions in the functions of memory. Every minute tracing of objects is held by the scroll of light, and no separation of message from messenger occurs; the pictures gathered through the pupil of the eyes are not lost but retained, not changed but stored; the pictures of the sun, moon or stars, drawn in the past remain for comparison with those seen to-day; the clouds in the sky now may be compared with those we have seen before; the forest, the orchards, the field and its fences, rivers, gardens and dwellings—all the countless objects of daily life have thrown more or less distinct pictures of facts into the gallery of memory, where they remain subject to inspection and reference unaltered and unconverted.

The pictures enrolled on the tablets of memory are not mere outlines of lights and shadows, not mere blendings of the spectrum of colors, not hyeroglyphics to be translated by transmutation, but contain as actual enrollments of light as steel does of magnetism, as water does of heat, or as muscles do of muscular force. That memory retains the forms and features of a tree, a house, a watch, people, and objects previously seen is entirely familiar and unquestioned. The feature here presented and insisted on is not as to the fact of memory, but that knowledge of objects seen is not manufactured, but gathered; not transmuted by the brain from impacting motions of substance, but that imponderable images drawn on impalpable light are as actual existences as imponderable chemical or magnetic force; that a sunbeam in its luminous and non-luminous parts is as entirely competent to carry and distribute information and knowledge to the senses of organic life, as heat is to carry motion and expansive force into inorganic substance.

In asserting this capacity of light to carry absolutely accurate information of surrounding objects to the retina, no issue is made with well known and familiar facts: the pictures are there formed on the nerve, and we know that they form an essential fact in the process of seeing, and we also know that we can recall the knowledge which they impart. After seeing a bird flying, a horse running, or a person standing in front of us, we can recall whether the bird was a blue bird or robin; whether the horse was large or small, black or white; whether the person was man or woman, young or old, friend or strange, dressed or undressed; and if the image formed on the retina was clear and distinct, our capacity to recall the minute tracings which light has effected are correspondingly minute.

After hours of repose and sleep, our eyes open and a glance around the room shows familiar objects; the ceiling, the door, the window and curtain of the room form pictures in the eye precisely like those in memory of the previous day; the forms and faces and features at the table form panoramic pictures for comparison with those already in the treasury of knowledge; we recall from memory the blooming health, the joyous sparkle which those around us were wont to exhibit, and compare them with the seared lines of care, the hectic flush of fever, or the gaunt spectre of death; we recall the budding leaf, the opening flower, the skipping lamb, and sprouting grass of spring, and follow in succession their changed phases through summer's growth and autumn's harvest, until the landscape is covered with a mantle of snow, and again repeat that these, and all pictures of things we see were traced and formed on the canvas of light, to pass into the vestibule of memory unaltered, where they exist in the gallery of facts unchanged, and are actually stored in the treasury of knowledge—pictures and light, panoramas and fact forming components of mind, grafted on the sentient nerves of life as heat can be grafted on oxide of hydrogen to form water—and that light has the power of holding images corresponding with its power of forming them. If we place ourselves in succession in all parts of a room, we can see in succession every minute part of the room, showing that light, with its pictures of facts, is traveling in all directions, crossing and re-crossing the lines of light from each and every object in every direction; but by virtue of the perfect transparency of the canvas and pictures, not the slightest interference in information from the different directions is produced; the crossing and re-crossing paths do not interfere with each other's accuracy of images on the retina. No matter how full the room or how small the object, each offers its representation to the retina for inspection and storage irrespective of images from other objects; and when they are entered on the tablets of memory the images from one object never obscure those from another.

That the pictured representations of facts and panoramas thrown into vision and memory are just what is retained, not only corresponds with our conceptions of pictures reproduced in memory, but the function is foreshadowed and explained by that property of light, which light has in common with all imponderable agents, of holding and retaining all their properties in each and all of their allotropic states and conditions. The pictures have gone through vacuums, gone through gases, entered the animal system through animal tissue, passed through the crystalline lens, passed through the humors of the eyeball, and have actually penetrated the animal system a considerable distance as absolute and actual pictures, and there is no known reason why they cannot remain in the substance of the organism with the imponderable agent that drew and carried them into the animal body.

That light, after penetrating the eyeball, remains and is not reflected out, is an ascertained fact; that it carries information into the mind, as pictured information, is familiar and evident from an inspection of images formed on the retina; that these pictures were traced on light is also evident, for we find the pictures on the messenger at any part of the route from the object to the eye; and that the panoramic pictures, or their exact counterpart, exists in memory, are all well known and familiar facts. Now when we look at these facts in conjunction with the fact of the evident presence of inactive dormant light existing in the apparatus for vision, the inference is very strong that the imponderable agent and pictures, panoramas and messenger, are mutual companions in the optic and sentient apparatus.

The simple fact that pressure on the eyeball causes a sensation of light naturally suggests that light is dormant in the apparatus of vision; then again if an electric current be passed through the eyeball or socket to the top of the head, flashes of light appear at each interruption of the current, and this fact confirms the inference that light exists in a dormant state somewhere in the electric circuit of the eye; then again if the optic nerve be cut, torn or injured strong flashes of light appear at each disturbance. In some cases of injury of the optic nerve a constant sensation of colored light appears in the eyes, a fact somewhat parallel to the ringing noises in the head from injury to the nerves of hearing.

Most persons, if placed in a dark room, will have a sensation of a very dim light close to their eyes when the lids are closed, and on opening the lids the dimness breaks into streaks, changing from place to place, but generally quite close to the eyes. Under these circumstances young and timid persons are very apt to strain their eyes trying to see through the darkness, and as the nerves become excited the light appears at a greater distance, and often flashing into spectral illusions of odd, or curious, or sometimes fearful shapes. In the phenomena of delirium tremens, the disordered nerves bring up pictures of coiling, hissing serpents, swarms of stinging hornets go buzzing and darting at all parts of his body, deadly enemies seem to rise in every direction, pass and repass only to make room for forms and shapes still more fearful. Similar effects occur in dreams, in an inferior degree; strange groupings of things and parts of objects once seen are intermixed in orderly or disorderly array; pictures, facts and thoughts are pleasantly or unpleasantly grouped in disjointed fragments, or in connected illusions. In the delirium of fevers strange phantasies, dimly pictured with incoherent flashes of thought rise up from memory.

By the mistake of a nurse that was attending me in a sickness, I once took at a single dose a quantity of quinine that was intended for twelve doses at intervals through six days. The medicine threw me into a delirium of half consciousness and partial sleep, and for about two hours a rambling running panoramic view of disconnected parts of all my previous life seemed to flit past my eyes in a disjointed jumble; all my boyhood's plays, pranks, struggles and scenes appeared; some of them as fresh as actual present sights and occurrences; the atmosphere seemed all aglow and illuminated with a fiery red light, but no hotter than usual; pictures of scenes one moment seemed to be away off in the distance, and then close by, sometimes as plain as an image in a looking-glass, then offering only a mere glimpse of fading illusions. I had been to a combined caravan and circus only a few days previous, and circus scenes were pictured on the running panoramas much oftener than any other event, but they were in a strange jumble; clouds in the fiery red sky would flash out in brilliant colors and tints, as are sometimes seen near sunset after a summer shower, and then fade into dim gray fogs; the show tent, with everything it contained of horses, wagons, people and animals, seemed to have been emptied out of the clouds and were performing just below in the air; elephants were turning somersaults with hyenas clapping their paws in admiration; tigers were performing on tight ropes swung from the clouds, sometimes hanging by their teeth or a single paw, or sliding down and holding to the rope by only a twist of their tails; horses stood on their heads and balanced clowns on their feet; babies sat in the lion's mouth and crowed and laughed at the frantic efforts of their mothers to get them; clowns, riders, animals and attendants were constantly shifting, changing and interchanging positions and roles, sometimes taking a natural, and then shifting into an unnatural position or part of the performance. This jumble of circus performances and animal show was continually shifting and giving place to pictures of other events, and then reappearing in disjointed fragments or bringing vivid, perfect and natural pictures. One moment I was a mere boy chasing butterflies and trying to catch them in my hat, then reciting lessons with my schoolmates, and again watching the daring feats of female equestriennes standing on their heads on flying bareback spotted circus horses.

Another effect of the quinine was to cause a sensation of ringing and various noises in my ears, some of them as loud as the report of a pistol, others only like a hissing, buzzing hum. The ringing noises did not disappear entirely for several months, and could be strongly excited and roused by pressing my head back of the ear. In a few hours after taking the quinine, the delirium and stupor passed off, but if I pressed lightly with my finger on the eye ball, pictures of various kinds would reappear, generally intermixed with disjointed fragments of circus scenes. For two or three days these panoramic pictures, developed from pressure, were clear and distinct, but the eyes gradually lost this susceptibility for developing pictures from pressure, so that at the end of a week, pressure on the eyeball would only cause flashes of light to appear without the pictures.

Dreams, deliriums, delusions and illusions of the imagination are well known to be associated with disordered nerves of the stomach, as well as the brain, and it is commonly taught that imagination and the entire range of mental phenomena grow out of molecular changes of brain structure and substance. That perception, and thought, and imagination, and all mental phenomena are the result of oxidations and molecular changes of brain substance has been asserted, and reasserted so often that a great many persons seem to forget that these assertions are only unproved theories, and accept them as ascertained and absolute facts; yet it is no more an ascertained fact that thought is due to combustion of nerve substance than it is that digestion is due to combustion of the stomach; no more an ascertained fact that imagination grows out of oxidation of nerve substance than it is that muscular force is founded on oxidation of muscles. The continued repetition of the assertion that thought is manufactured in the brain by a process of chemical change of brain substance, has fixed the idea so firmly in the minds of many persons that to call the notion an unproved theory will seem like sacrilege; yet these assertions, so frequently made, that thought is brain work, produced by wearing out nerve matter is wholly and entirely theoretical. The chemical whirligig of the molecules of brain substance takes place much more rapidly after death than during life, and no one would have the audacity to assert the production of perception, imagination, or thought, from the decomposition that takes place in the grave. The chemical whirligig of brain molecules in organic decomposition may be retarded and made to proceed rapidly or slowly, but their modification does not develop or produce mental phenomena. This distinction between theory and fact, between what we know and what is currently taught and believed, is essential to an understanding of the explanation of dreams, illusions, memory, or mental phenomena that is here presented; for if we discard a hypothesis merely because it conflicts with some other hypothesis that we have learned, we are just as liable to discard truth as error.

The explanation which is here offered of memory, imagination and illusions, does not tally with the oxidation theory of brain work; but the difficulties that spring up and gather around the attempt to attribute the production of pictures from the tablets of memory to oxidations of brain substance seem insurmountable.

Although it is often broadly asserted that the imagination is the result of chemical changes in the brain substance, these changes have never been discovered, nor even conceived in any definite sense. The adjustment of just the suitable degree of oxidation of nervous matter in millions of different brains, in all their varying conditions, that would enable each and every one to mutually and correctly observe at the same time a person, a landscape, a circus, or other panorama of facts, and the adjustment of just the suitable degree of oxidation of brain substance to reproduce in imagination the same panorama as of circus, landscape, or person at different times and at optional intervals of days, months, or years, is an absurdity so great that no sane mind would entertain it for a moment, yet the oxidation theory of mental phenomena has no better foundation than this indefinite conjecture.

In discussing the mental phenomena of dreams, delusions, illusions, imagination or thought, the same posttion, is presented that was presented in discussing animal heat, vital and muscular force. That is, that mental facts are gathered in the same sense that motor forces are gathered by their respective apparatus; that the motor forces of heat, or muscular force are merely gathered—not generated—as new existences; that they already exist, and the vital processes of animal life merely gathers and distributes them— some of the processes bringing muscular force into the muscular apparatus, where it is stored for use. And so of mental facts and impressions, derived through the senses, they are absolute existences and not manufactured by the brain; are carried into the mental apparatus by means of both illuminating and non-illuminating rays of light through the senses.

It is a familiar fact that animal food must consist of substance holding vital force in union with the substance that embodies it as an essential part of its substance. Carbon and hydrogen, and all the material elements found in a live potato by the chemist, may be found in one that has been frozen and thawed to rottenness; but after its vital force is eliminated by rotting, it is utterly incompetent to furnish vitality to the living animal. The living animal is utterly incompetent to manufacture vitality from substance deprived of it; it can only distribute it with substance wherein it already exists.

The animal system never generates or produces vital force in any other sense than to extract and distribute what is already in existence. And so of animal heat; it already exists in the dormant state in the food we eat, the water we drink, the air we breath, and the various processes for its production in animal life are simply processes for developing from it its dormant condition, and not processes of manufacture. An oxidized substance, like ice, that contains no absorbed heat, is utterly incompetent to furnish heat to animal life; the animal, in any of its processes, can only extract heat, or warmth, from substance to distribute to its various parts. Breathing, digesting, oxidizing, eliminating, or any other animal process, never generated or produced a fraction of heat, in any other sense than to develop and guide what already existed, in substance, in, or within the animal body. So, too, of muscular force, a limited amount is carried into and stored in the muscular apparatus that can be more rapidly exhausted than replenished by the processes, and when this occurs, the animal must wait for a new supply to be extracted from its nutriment and stored in the apparatus.

This principle has already been discussed in its relation to the so-called motor forces of heat, muscular, and of vital growth—forces that produce the various movements of organic substance—and is referred to here for the purpose of again pointing out that the same principle which underlies mental knowledge, underlies the motor forces of life. Facts are gathered and stored in the mental apparatus, and whether recalled and reviewed in natural order and position, or whether they rise up in disjointed and distorted illusions of the imagination, they are not manufactured, not transmutations of thrills or shimmer of nerves, not undulations of brain matter, but are actual existences drawn on the canvas of light, and exist in the mental storehouse. The education of the machinist brings to his optic nerve pictures of wheels, cranks, cogs, links, levers, weights, screws, bolts, plates, shafts, rods, boxes, pumps, and all the different paraphernalia of machinery that are constructed of wood, metal or other material, in all their varying dimensions, and these pictures remain in the storehouse of memory to be called up in their several parts to be arranged, or rearranged in different dimensions, in different relative positions, in different material, in adaptation to varying circumstances required in practical life, or these pictures may rise in disjointed and distorted fragments in the delirium of abnormal processes.

The writer, in writing, recalls in succession the forms of the different letters that spells the different words of his sentences, and what is here asserted in relation to it is, that these forms of letters actually exist in the world of facts, and that they are gathered and carried into the apparatus of vision, spread out, or diminished on the optic nerve as facts, and may be tinted of any and all the various colors, or may be colorless. If these letters were so tinted with all the various colors, the inconceivable complication of red, blue, violet, black, white and yellow would give such a complicated system of undulations of the optic nerve to unravel, decipher, or translate into knowledge, that it would be difficult to understand how forms could reappear by the process of undulations through their infinite motions.

The artist reproduces with still greater fidelity the features, forms and faces of those he may wish to represent on canvas. His representation of an apple, peach, cherry, or other tree with its fruit; a rose, lilly, or other flower; a negro, Chinaman, Indian, or other object may be represented, large or small, with or without color; and if their forms are in accurate proportions they are immediately recognized when there are corresponding pictures in the storehouse of memory for comparison.

The caricaturist delights in reproducing from his storehouse of pictures, inaccurate forms, by wrong and absurd groupings of the different parts of his memory pictures; very large heads, on very small bodies, or very large bodies with very small heads and very short legs, spindle shanks, huge noses, bleared eyes, uncouthly dressed figures, and the infinite jumble of disarranged parts of the various things and phenomena that have passed across his vision. The germ, the bud, the full grown organism in all its various stages of growth, are placed in the storehouse of memory; the entire material world, in miniature, or their forms in mass, in whole, or in parts, are constantly offering pictures through the apparatus of vision. These pictures are grouped, and regrouped in mental process, but whether recalled in parts, or in whole, accurate or inaccurate, definite or vague, their forms are described on the messenger of light; information and fact, with the messenger of knowledge, pass into the organ of vision, to be stored as stones will store weight, steel will store magnetism, or as mercury will imbibe and store heat, or a glass jar will store electricity. Apprehension and discernment, knowledge and fact are as absolute as force, and may be stored in sentient apparatus with as absolute certainty as chemical affinity can be stored in chlorine or potassium, the catalyzing action in platinum, or the actinic force in light.

Sound and Hearing Due to Non-luminous Conditions of Light.

The senses are sometimes called the windows of knowledge, and much the greater part of our actual knowledge is derived through these windows, or senses. To assert that all information of the qualities and properties of bodies—all facts that we call knowledge—are gathered on certain conditions of light, and reach consciousness on this messenger, may at first appear a very broad assertion; yet this is just what is here presented. Information and knowledge not only reaches the eye on the canvas of light, but flavors, and odors, and feeling, and sounds are each carried by conditions of the same medium.

To assert that knowledge, derived through the sense of hearing; to assert that conversation, music—all sounds—are carried on a condition of light, is to make statements that conflict with accepted theory, but is, nevertheless, true. Each of the five senses brings to our perception a distinct class of information; the sense of seeing discerns the superficial qualities of objects; the sense of hearing discerns the inner qualities of objects revealed by sounds.

The mode in which sounds impart information, as commonly taught, is thus stated by Professor Tyndall: "Sounds we know to be due to vibratory motion."......" What is sound within us is, is outside of us, a motion of air."

"From the earliest ages the questions, 'What is light?' and, "What is heat?" have occurred to the minds of men, but these questions never would have been answered had they not been preceded by the question, 'What is sound?' Amid the grosser phenomena of acoustics the mind was first disciplined, conceptions being thus obtained from direct observation, which were afterwards applied to phenomena of a character far too subtle to be observed directly.

"Sound we know to be due to vibratory motion. A vibrating tuning fork, for example, moulds the air around it into undulations of waves, which speed away on all sides with a certain measured velocity, impinge upon the drum of the ear, shake the auditory nerve, and awake in the brain the sensation of sound. When sufficiently near a sounding body we can feel the vibrations of the air. A deaf man, for example, plunging his hand into a bell when it is sounding, feels through the common nerves of his body those tremors which when imparted to the nerves of healthy ears, are translated into sound.

There are various ways of rendering these sonorous vibrations, not only tangible but visible; and it was not until numberless experiments of this kind had been executed that the scientific investigator abandoned himself wholly, and without a shadow of misgiving, to the conviction that what is sound within us is, is outside of us, a motion of air. But once having established this fact—once having proved beyond all doubt that the sensation of sound is produced by an agitation of the nerve of the ear, the thought soon suggested itself that light might be due to an agitation of the nerve of the eye. This was a great step in advance of that ancient notion which regarded light as something emitted by the eye, and not as anything imparted to it. But if light be produced by an agitation of the optic nerve or retina, what is it that produces the agitation? Newton, you know, supposed minute particles to be shot through the humors of the eye against the retina, which he supposed to hang like a target at the back of the eye. The impact of these particles against the target, Newton believed to be the cause of light. But Newton's notion has not held its ground, being entirely driven from the field by the more wonderful and far more philosophical notion that light, like sound, is the product of wave motion. The domain in which this motion of light is carried on lies entirely beyond the reach of our senses. The waves of light require a medium for their formation and propagation; but we cannot see, or feel, or taste, or smell this medium. How, then, has its existence been established? By showing that by the assumption of this wonderful intangible æther, all the phenomena of optics are accounted for with a fullness, and clearness, and conclusiveness, which leave no desire of the intellect unsatisfied."—*Tyndall on Radiant Heat and its Relations in Fragments of Science.*

"The ether which conveys the pulses of light and heat not only fills celestial space, swarthing suns, and planets and moons, but it also encircles the atoms of which these bodies are composed. It is the motions of these atoms, and not that of any sensible parts of bodies that the æther conveys.

It is this motion that constitutes the objective cause of what in our sensations are light and heat; an atom then sending its pulses through the æther, resembling a tuning-fork, sending its pulses through the air."—*Ibid.*

"Thus far we have fixed our attention on atoms and molecules in a state of vibrations, and surrounded by a medium which accepts their vibration and transmits them through space. But suppose the waves generated by one system of molecules, to impinge upon another system, how will the waves be effected? Will they be stopped, or will they be permitted to pass? Will they transfer their motion to the molecules on which they impinge, or will they glide around the molecule, through the intermolecular spaces, and thus escape? The answer to this question depends upon a condition which may be beautifully exemplified by an experiment on sound. These two tuning-forks are tuned absolutely alike. They vibrate with the same rapidity, and, mounted thus upon their resonant cases, you hear them loudly sounding the same musical note. Stopping one of the forks, I throw the other into strong vibrations, and bring that other near the silent fork, but not into contact with it. Allowing them to continue in this position for four or five seconds, and then stopping the vibrating fork, the sound has not ceased. The second fork has taken up the vibration of its neighbor, and is now sounding in its turn. Dismounting one of the forks, and permitting the other to remain upon its stand, I throw the dismounted fork into strong vibration. You cannot hear it sound. Detached from its stand the amount of vibration motion which it can communicate to the air is too small to be sensible at any distance. When the dismounted fork is brought close to the mounted one, but not into actual contact with it, out of the silence rises a mellow sound."

"Whence comes it? From the vibrations which have been transferred from the dismounted fork to the mounted one. That the motion should thus transfer itself, through the air it is necessary that the two forks should be in perfect unison. If a morsel of wax not larger than a pea be placed on one of the forks it is rendered thereby powerless to affect, or be affected by the other."—*Ibid.*

Professor Tyndall has long been known as a prominent advocate of undulatory theories. In the above quotation we have a clear and concise statement of what these theories are supposed to explain. According to the view here put forth, a complete and perfect explanation of the phenomena of hearing and sound is a matter of easy observation. Sound is motion, vibrations of matter, "tremors, which when imparted to healthy ears, are translated into sound." Previous to this answer to the question, "what is sound?" no adequate answer to the questions, "what is light?" and, "what is heat?" could have been given. But after the mind had been disciplined in the study of, "what is sound?" and had obtained the full and complete answer, the answer to this question furnished a key that unlocked the more intricate mysteries of sunbeam, and led to a knowledge of what light and heat are.

The key that unlocks the wonders of these marvellous agents is simply, "vibrating motion." In the impact of atoms, and the crash of worlds, the Professor sees a sufficient cause to explain the phenomena of sound, light, heat, and the evolution of organic forces. In his language, "sound is known to be due to vibrating motion." "To this fact the scientific investigator abandons himself wholly without a shadow of misgiving to the conviction that what is sound within us, is outside of us a motion of air."

Once having established, to his own satisfaction, that vibrating motion is the sole condition of matter, in the sense of hearing, it was very easy to conjecture that the senses of vision, smell and taste, were also dependent on vibrations imparted to their respective nerves; but as no vibrations for these senses, could be observed or detected, to transmit light and heat, another conjecture was required to make their explanation tally with what is claimed to be a full explanation of sound. This conjecture gives us an undiscoverable substance enveloping atoms and worlds— the undiscovered "æther." Before accepting vibrating motion, as the master key to unlock these mysterious agents, let us see if it unlocks all the phenomena of sound.

UNDULATORY THEORIES.

The history of the past shows that notions and theories which are accepted at one period, are afterwards modified, changed or discarded. At one time it was taught that heat was wholly dependent on combustion; the sun's heat, the heat of fires, chemical action, animal heat, all seemed to corroborate this view; but as the heat developed by friction, by percussion, electricity and magnetism was studied, it was found that the combustion theory was too narrow to embrace all modes in which it was developed. So also of electricity. At one time it was conjectured that as a galvanic current was dependent on chemical action, electricity was developed in this way in all other cases; but when it was discovered that a current could be developed by heating two metals, by twisting wire, by friction, by magnetism, and various other modes, it was conceded that the theory of chemical action was entirely too narrow, and did not embrace all modes for its development. So of the sensation of hearing. If the sensation of sound is, or can be produced in any other mode than by vibratory motions, then the theory that sound is wholly due to vibrating motion is also too narrow; for a theory that ignores or fails to embrace all facts is an insufficient explanation.

This is precisely the standing of the undulatory theory of sound. It ignores, or fails to explain why people hear, or seem to hear various noises when no vibratory motion can be traced. Vibrations or undulations of air, and the sensation of hearing are simultaneous facts in a large class of apparent sounds; but the sensation of hearing is also quite common in which no vibrations, or undulatory motions of matter has been traced; and to assert that hearing is wholly due to vibrations, is to ignore this very large class of phenomena. Thousands and thousands of people in all parts of the world experience a sensation of hearing, a ringing or buzzing sensation in their ears when no sounds are to be heard by others; others hear voices, music, guns, and various sounds that have no foundation outside of themselves; yet these sounds are as real to them as to those that are carried by atmospheric vibrations. These sounds are induced by injuries to the head, or nerves, by medicine, by overtaxing the stomach, sometimes by compressed air, and by other modes, showing that vibrations of matter is not the only mode in which the sensation can be developed or produced.

Then again there are other modes in which sound is propagated, besides the vibrating mode. Those who claim that sound is merely vibrating motion, assert that when these oscillations reach the nerve of hearing, they are translated into sound. In the language of Tyndall, "what is sound within us is, outside of us a motion of air, tremors which are translated into sound."

By surrounding atoms and worlds with an inpalpable æther to catch and transfer a quivering motion, a quite prevalent notion has obtained in the human mind—a notion sometimes expressed, and sometimes merely implied—that the phenomena of the universe may be explained without the assistance of what are known as imponderable agents; by the aid of an æther swarthing suns, planets, and atoms, light and heat are only a quivering of æther—sound merely a quivering of air. This last dogma, in explanation of sound, has been accepted so largely as positive fact, by the public, that to question its accuracy and competence as a full explanation, and suggest that sound is something more than mere vibrating motion, will, to some, seem as a useless waste of time and paper; yet there is probably no dogma ever enunciated by scientists that has had a more pernicious influence on human thought than is contained in the dogma that sound is merely vibratory motion of air. It is the corner-stone and foundation of undulatory theories—theories that attempt to banish from the universe all the intermediate agents, like light and heat, between omniscience and phenomena; theories that would not only banish light, and heat, and immaterial forces of the inorganic world, but place sensation, thought, knowledge, and life among the vanishing quivers of vibratory motions; deny them position as real existences, and substitute for them the mere quivering throb of atoms.

In questioning the fullness and competence of the vibratory theory of sound to explain all its phenomena, no issue is made with ascertained and known facts. There is no question but that a sounding bell, or other body, vibrates or that these vibrations are imparted to the surrounding air; but the question at issue is, whether only motion is propagated? By the view here presented, the vibrating state of matter, emitting sound, is simply a condition of matter enabling a subtle agent to pass, in the same sense that the transparent condition of matter enables the subtle light to pass—precisely in the same sense that a heated plate of iron enables heat to pass through it.

The vibrating bell, when struck with a hammer, undoubtedly propagates sound; and by the view here presented, to do this, there is developed and propagated to the quivering air, an imponderable agent, in the same sense that heat, as an imponderable agent, is also developed and propagated by the hammer's blow. Both of these agents are developed by the blow, and propagated to surrounding bodies, not as motions, but as actual, indestructible existences. These two agents, the agent of heat, and the agent of sound, which can thus be developed by the hammer's blow, are twin sisters in a sunbeam.

At the present day it is generally conceded that all the heat and light existing, or that can be developed on the earth, by combustion or other modes, originally came from the sun.

In tracing other modes in which sound is transmitted or propagated, we find the telephone wire taking up all classes of sounds and carrying them in conjunction, associated with each and all the royal family of imponderable agents. In this simple and well known fact, that sounds are taken up by an electrical current, propagated and transmitted with electricity, heat, magnetism, light, and chemical affinity, over vast distances, along conducting wires and through the ·earth circuit without a perceptible tremor, and without the slightest trace of vibration, the inference is natural, if not unavoidable, that the agent of sound, and battery agents, have a common origin and a common destiny. Until the fact that sound is transmitted or propagated by an electric current, without a perceptible tremor, is overthrown or explained by arguments more powerful than conjectured motions of the atoms of telephone wires, the assumption that sound is mere vibratory motions of matter, stands as a theoretical dogma and not an ascertained fact; and before the undulatory theory of sound can consistently be extended into an imaginary æther to explain the phenomena of vision, taste, smell or feeling, it must have a better foundation than the mere simultaneous facts of vibrating air, sound, and hearing, for this simultaneousness does not establish that hearing is dependent on vibration.

The subtle agent, light, is the grand carrier agent to impart information and knowledge of the qualities of different objects; and that part, or condition, of light adapted to the sense of hearing carries information of the inner qualities of bodies. If a bar of steel be struck with a hammer it emits a sound, that, to the practiced ear, gives as accurate information of its inner qualities as illuminating light would of its superficial qualities. A person holding a bar of steel, a bar of rolled iron, and a bar of cast iron, and striking them with a hammer, will readily distinguish the different sounds emitted from each. Workers in steel will distinguish in this way the different qualities of cast, blister, and spring steel, and will also distinguish and detect flaws, and impurities, and determine fine from the coarser varieties. The wide difference between the harsh rattling twang of Chinese gong metal, and the muffled, deadened tone of common rolled brass, when struck with a hammer, is detected by an inexperienced ear. The vibrating conditions of the different metals propagates an agent that reveals the precise condition of their inner qualities. The footsteps of a common house fly, in its ordinary walks, is never heard; but as it walks on the sounding board of a microphone, the sound of its footsteps is taken up by the electric current, without the slightest traceable vibration, and carried to a distant part of the circuit, and there repeated.

In explaining the fact that sound can be, and is, carried by electrical currents without vibrations, I assume that the electric current is a group of imponderable agents, and that *light* is one of the group. This fact, that light is one of the members forming the group of agents of an electrical current, will be referred to again in noticing the influence of a current on nerves, and organic matter and life.

The group of agents in a sunbeam, when separated, reveal heat as a distinct agent, chemical influence as a distinct influence or agent, and light as another distinct agent, or part of the beam. Each of these distinct agents of the sunbeam, as is well known, have distinct properties which become manifest under varying conditions. Thus heat, as it exists in the sunbeam, readily passes through the intervening space between the sun and the earth; readily passes through glass and other transparent bodies; but if we examine obscure heat of low temperatures, as for example, heat from a plate of metal heated in boiling water, we find that the heat emitted from the metal will not pass through glass, or through a vacuum. Although this heat is the same heat that originally came from the sun and has lost no energy, yet we here find it manifesting different properties from heat as it existed in the beam of combined agents from the sun. These different properties of heat and other agents are what I have termed allotropic conditions. In the one condition we find heat and light traversing the distance from the sun in less than ten minutes; in another of these conditions we find both heat and light dormant and asleep; in one condition traveling over ten million of miles a minute; in the other condition perfectly torpid, and without apparent energy. In its allotropic condition, as contained in or propagated from the heated metal or hot water, heat manifests a comparatively sluggish energy, as compared with its original energy and motion. The phenomena of dormant, torpid heat, in ice water, has become quite familiar to students—a condition in which its ordinary properties are obscured.

Similar facts obtain with light. We commonly know of light as the swift messenger of knowledge, bringing information to the eye of distant objects—of their superficial qualities. It is also the messenger to each of the other senses; it is the messenger that carries information of the inner qualities of steel, brass, wood, and other bodies which, when struck, are made to vibrate. Light, like heat, has its several allotropic conditions; it exists in the torpid, dormant condition, fast asleep. It is in this condition when the bell, tuning-fork, or other body is made to vibrate. The vibrations develop it from its state of torpor, and becoming partially roused from its lethargy, it creeps along the vibrating substance; but, if allowed to cling and ride with other members of its family in an electric current, its motion is more rapid, confined to narrower lines it moves with greater velocity.

Low tension heat, that refuses to radiate through a plate of glass, is called obscure heat, while the entire beam from the sun, including its heat, is commonly called light; but much the greater part of this beam is of a non-illuminating character. The entire beam, however, is light, in the sense of being a medium and messenger of intelligence. In this sense the heating part of the beam is also light. We are conscious by the sense of feeling its warmth, of the presence of a warm body, and heat, emitted from the warm body, imparts the information. To do this the body first absorbs heat before imparting or emitting it; and as we have already noticed the absorbed heat sometimes remains totally obscure and dormant for indefinite periods of time before being emitted. And so of that part of the beam that affects the sense of hearing as sound; it is absorbed, obscured, and remains dormant for indefinite periods of time, ready to be propagated by the vibrating conditions of matter. This competence, or property of passing into the dormant, inactive state, enables the agent to pass into the dormant, inactive state in the organs of hearing, to be recalled for comparison in the process of remembering sounds once heard.

But how it is possible for mechanical vibrations to pass into the dormant state and afterwards to become roused as identically the same vibrations that vanished from existence, is an unsolved mystery and unsolvable problem.

Heat that imparts the sensation of warmth is propagated through matter and adheres to that quality of matter that undergoes expansion; and it is heat emitted from expanded bodies that carries information of their temperature.

If we attempt to apply undulatory theories in explanation of the phenomena of human or organic life we soon find ourselves drifting into a jungle of impenetrable darkness—a vast ocean of empty nothingness. Mankind are provided with means for producing and maintaining an equitable temperature of 100 F. through their life period, and if this function is interfered with, by which higher or lower temperatures obtain control, the entire system becomes deranged and life endangered. The undulatory theory asserts that heat is vibratory motion—a quivering of the universal æther, and that this æther permeates the animal body, surrounding its atoms and molecules, and that every minute part of this permeating æther vibrates and quivers many millions of times each second to maintain the warmth of a living human body—the vibrations of the æther being taken up by their material substance and constituting what we term warmth, or animal heat—it being understood in this connection that cold hands, feet, or other parts receive a fewer number of these mechanical impulses than the warm vital parts.

How the sudden variations of temperature—cold chills, and sudden flashes of fever from one part of the system to another, are to be brought on by the all-pervading æther, is a puzzle and an unsolved problem. The warm nerves of hearing are supposed to be in this constant state of vibration like the rest of the body, to maintain their temperature.

The senses of animal life are not only means for receiving impressions and sensations, but they also have, or are associated with the faculty of remembering that similar sensations have been received in the past. The ear hears the roaring of lions, growling of tigers, braying of asses, barking of dogs, chirp of crickets, the ticking of a watch, or the sharp jingle of cymbals, the ripple and babble of brooks and the patter of rain drops, the whirling of winds, and the screech of panthers, the voice of command and sobs of entreaty, the round rippling laugh indicative of joy and the cadence of sorrow, the wild jar of discord and the concords of music—all the intricate variety of untold billions of different sounds reach the ear and give information of the objects that produced them. Sounds once heard are immediately recognized if heard again years afterwards; but in what way these sounds, if they are merely undulations, merely vibrations of air, can be recalled for comparison after having totally vanished, has never been explained and appears unexplainable.

The undulatory theory of sound brings to the nerves of hearing a series of undulations vastly different from the undulations of æther that produces (in conjecture) the animal warmth. The vibrations to produce sounds only range from sixteen undulations up to sixty thousand undulations a second, while the æther, to produce heat, undergoes millions and billions of quivers. In the experiment quoted from Tyndall, the two tuning-forks are timed absolutely alike, so that when one is set vibrating and the other brought near it, the still fork catches up the tone and vibrates in unison; that it shall do this, however, the two forks must be tuned absolutely alike.

"That the motion should thus transfer itself through the air it is necessary that the two forks should be in perfect unison. If a morsel of wax not larger than a pea be placed on one of the forks it is rendered thereby powerless to affect, or be affected by the other."

The fork containing the morsel of wax or other additional matter refuses to receive the tones of its former mate. If this principle obtains in the nerves of animals, how can the nerves of different animals, or the nerves of the same animal at different periods of its life, receive the same tones? Some are large, some small, some young, some old, some lean, some fat, and hence joined with varying quantities of matter. The nerves are also already supposed to be vibrating with a velocity beyond conception to maintain animal warmth; and how, with these motions, are they to receive, at the same time, the slow oscillations of sound, the varying and intense vibrations for vision, smell, taste, and feeling, unless we discard the principle of unison set forth by the tuning-forks and conjecture some different principle? This, however, is only the commencement of the difficulty. Each of the other senses are, by the undulatory theory, supposed to receive and carry by their respective nerves, other, and different rates of vibrations to be transmuted, or translated into mental impressions.

The undulatory theory asserts that "what is sound within us is, outside of us, motion of air;" that the ticking of a watch, chirp of crickets, song of birds, music of bands, voices of people are, before they reach the brain, merely the vibrating motions of matter; that the vibrating atoms of the bell or trumpet jog the atoms of the atmosphere with a mechanical impulse; the atmosphere jogs the nerves of hearing with a similar impulse, the nerves of hearing jog the brain, the molecules of the brain jog these motions of air and nerves, and carry these mechanical impulses and translate them into sounds and conscious knowledge.

The undulatory theory of light asserts that incandescent combustion jogs the substance of the universal æther, the æther in its vibrating impulses jogs the surfaces of objects, and these, in turn, jog the æther, the jogging æther jogs the nerves of vision, these jogging hints jog the brain and these "strictly mechanical impulses," the molecules of the brain, translate into actual knowledge of objects. We direct our eyes to a tree loaded with leaves and fruit; each of these thousands of leaves is different from its fellow leaf, and each of their analogies, and each of their differences, sends its vibratory hints to the brain for translation. Each leaf is a little larger, or a little smaller, its point a little more sharp or a little more blunt, a little more curled or a little more straight, turns more to the right or more to the left, a little more up or a little more down, its edge is a little more notched, or a little more plain, its ribs more plump or more slim, it has a little more green, or a little more yellow; having these, and a thousand more minute differences, from its fellow leaf, yet each and all these minute differences are supposed to be sent to the brain for explanation and translation by the infinitesimal quiverings of æther; these qualities and differences jogging the nerves, the nerves jogging the brain by such varying rates in vibrations that the brain understands and conscious knowledge emerges from the hints. The various tones of a band of music reach the ear, and we discern a national air; from drum and fife, trumpet and horn, clarionet and cymbal, cornet and flute, the high mellow voice of woman, and the full round bass of man, each and all send to the ear the same air in different tones, not as sounds, but as vibratory hints to be translated into sounds—into song and music.

The millions of pulsatory hints sent through the ear, and the billions on billions of pulsatory hints sent through the eye, are not the only hints to be translated into mental facts; each of the other senses are also sending a stream of hints to the brain. If undulatory theories are correct there is a constant stream of vibrations to the ear for each tone, thousands for each different scent, millions from all parts of the body as hints of feeling, billions to the sense of vision, and all knocking at the brain for translation into the five classes of knowledge discerned by the senses.

Ever since the undulatory theory of sound was adopted

there seems to have been a vague notion, in the minds of many writers of note, that imponderable agents might be discarded from philosophy, and effects previously referred to these agents explained by a conjectured æther undergoing various motions. Under this assumption, the notion, either expressed or implied, obtains, that the successive impact of atoms on atoms, or matter on matter, presents the sum total of cause and effect—the non-ending circle of the entire phenomena of atoms and worlds. But suppose the brain is jogged by merely mechanical hints pulsating through the several organs of sensation by æther and air, how does knowledge emerge from this crash of atoms and impact of molecules? The change from motion to thought is too vast for us to conceive. That motion generates thought, and to assume that what emerges from the brain is still mere motion, is to stultify our knowledge of mental phenomena—to flatter ourselves that we have explained them.

The transcendent sensation of hearing was not explained, nor the key to sensation found by tracing the simultaneous facts of vibrations, sound, and hearing. There is something besides pulsation of air to reach the sense of hearing; the qualities imparted to sound, by which we instinctively become conscious of the inner qualities of bodies that emit them, are not propagated by mere motion. The sensation of dread induced by a serpent's hiss, of fear from a lion's roar, is as certain and accurate a discernment of their dangerous qualities and their relation to man, as are carried by the images of their forms formed in the eye. What is here assumed and asserted is, that the same messenger that travels through transparent bodies and reveals outward forms in the one case, travels, or is propagated through the vibratory condition of matter in the other case. In both cases, knowledge of distant objects is received—not manufactured from mechanical hints, but is enrolled on the canvas of light. Knowledge brought to the sense of hearing by the messenger of non-luminous light, enters the storehouse of knowledge both for immediate and for future use. Both canvas and enrollment, message and messenger, pass into the allotropic condition of inaction, from which they may be readily roused in the process of memory.

The phenomena of sound is not transmuted nor translated motions, but actual fact. The sounds that we hear are just what the sense of hearing declares them to be, and just where the sense of hearing locates them. To assert that sound is only in the brain, is to discard all of our actual knowledge derived through this sense and to substitute an erroneous theory. We hear the rolling, reverberating thunder in the clouds, and that is where it is produced. Knowledge is as absolute as fact; and the messenger that brings and reveals to sentient existence, facts, as declared by each of the five senses, reveals them as enrolled knowledge; the agent that declares to the eye the existence of a distant ship, man or tree, embodies actual knowledge, and not quivering hints to be translated into knowledge by molecular changes of the brain substance. The embodiment of light as an essential part of man is analogous, and no more strange than the embodiment of heat as an essential agent of man and animal life. That light should enrol knowledge is entirely analogous, and no more strange than that heat should embody and carry force—no more strange than that an electric current should embody or carry sound, or magnetism, or heat, on connecting wires. What is here asserted of light embodying knowledge received by the eye, and knowledge received by the ear, applies to each of the other senses; it is not only true that the eye receives actual knowledge on the canvas of light, true that the ear receives its knowledge by the same messenger, but it is also true that each of the senses of smell, taste and feeling, receive actual knowledge of facts, enrolled on the same agent—a knowledge that is just what each respective sense declares.

What possible advantage can be derived by the brain from this storm of inconceivable motions, this impact of atoms and molecules on nerve or brain substance as set forth by undulatory theories? None of these theoretical quivering hints through the nerves have ever been discovered. If they reach the brain they vanish in a vast ocean of nothingness—leaving the mind that attempts to follow them in a confused jungle of absolute darkness that ends in an eternity of unceasing night. This entire system of jogging hints is largely founded on the original observation that sound is accompanied with vibrations, from which sprung the theoretical assumption that all that affects the nerves of hearing is simply these vibrations.

By keeping in view the fact that what touches consciousness in sensation is by the current theory merely conjectured, the relation of theory to theory turns on capacity of explanation.

According to the teachings of astronomy the sun is shining from a disc eight hundred and eighty-seven thousand miles in diameter. Theories of light teach that every part of a shining, or illuminated body emits rays in all directions. By this view, rays from each and every part of the sun's surface pass through the narrow pupil of the eye when the sun is seen. If we conceive that this immense surface is covered with fine luminous grains of sand, and that from each and every grain of sand there is a spider's thread attenuated to the smallest conceivable size, and one of these threads from every grain of sand passes through the pupil of not only one eye, but of every eye, looking at the sun, one end of the thread touching the uncounted billions of sand grains and the other end touching the retina, then conceive each and every thread vibrating in every forty-thousandth part of its length not less than five hundred billions of times a second—if these immensities are conceivable—we can have a faint notion of the load under which the undulatory theory of light staggers. This theory is frequently called by scientific writers the grandest conception ever made by the human mind. If incomprehensibility is essential to grandeur, this theory should be accorded the front rank. If these immensities of luminous surface and undulatory shimmerings are comprehended— if they ever are comprehended—we are no nearer a conception of what constitutes vision than before the mind was loaded with them. It seems like an attempt to account for the unaccountable—to explain the unexplainable.

The theory is here noticed because many leading writers seem to have adopted the notion that a theory of light must be presented; but why we should try to account for the existence of light and heat and other imponderable agents any more than we should explain and account for the existence of gold or iron is inconceivable. All such attempts heretofore made, or that probably ever will be made, only adds mystery to impenetrable darkness. We can examine the qualities and properties of ponderable matter, trace in succession the influence which imponderable agents exert in changing the forms and positions of ponderable matter, learn the laws regulating the relations of ponderable and imponderable existences, but beyond this no human mind can go—the inner essence of each is as unfathomable as infinite space.

The apparatus of the ear sustains the same relation to the sense of hearing and to the messenger of non-luminous light with its enrollment of knowledge of the various qualities of sounding bodies, that the apparatus of the eye does to luminous, or illuminating light with its enrollment of knowledge in images. The enrollment of Yankee Doodle, Hail Columbia, Home Sweet Home, on the agent of sound—an agent that traverses the vibrating atmosphere—is as actual as the enrollment of the forms of a horse, a house, or other body depicted on the retina of the eye.

The organs of hearing differs in different classes of animals. as is well known. An oyster's organ of hearing has not, with absolute certainty, been discovered; but that they hear is well known. When quite a small lad I passed by a table in an outhouse, on which there was a number of these animals in their shells. I shut the door with a slamming noise and was very much surprised to hear their shells close with a slamming force as they withdrew into their shells. The organs of hearing in the fish are covered with a semi-transparent membrane that prevents air or water entering into the labyrinth and jogging the nerve with its pulsations, but it is well known that their sense of hearing is quite acute.

That light is the enrolling agent that is propagated to impart the sensation of sound is very naturally suggested by the phenomena of singing flames. A common form of these flames is produced by placing a glass tube about thirty inches long. and two or three inches in diameter, over a gas flame; the tubes are sometimes made of metal, sometimes of wood, and as has already been noticed, tuning-forks, and other bodies will vibrate in unison with certain tones—certain sounds inducing vibrations, that in turn emit the same tone. On imparting the proper tone to the tube surrounding the flame it will emit the sound, and this will be taken up by the flame and both flame and tube will continue the tones indefinitely, unless disturbed. The vibrations of the air and the gases of the flame have been carefully counted and estimated and the number of vibrations for different tones ascertained. Tubes made of wood, metal, glass, or other material of different sizes and lengths, all differ in tones; and what is insisted on here is, that the non-luminous part or conditions of the rays emitted are competent to produce this phenomena of absorbing, emitting, and carrying sound from flames.

A tinman's ordinary soldering iron, when heated below redness and before it emits light, if laid on a cold block of lead or smith's anvil, will emit a singing noise of a nature like singing gas flames. Now if the diaphragm of a telephone be set vibrating by either of these or other sounds, the insulated wire around its magnet catches up the carrying agent with its information of quality in tone, and carries them without vibrations for miles, and then delivers the agent and its information of song, music or noise, to the sense of hearing, showing that the carrying agent that reaches consciousness in hearing assimilates perfectly with what has long been known as imponderable agents. This evident analogy of sound to imponderable agents has been traced in a variety of ways, and has had a strong influence in attempts to bring other imponderable agents down to, and explain them by mere vibrations. But in these facts there is evidently something more subtile than vibrations of matter; for the induction wire around the telephone magnet may be insulated with glass, gutta-percha, or by a vacuum, and still the wire catches, and carries, and repeats the sounds; and to suppose that sounds can be carried through vacuums by vibrations is to discard the current explanation of sound.

This fact, that sounds, and the agent that carries them, can be made or guided to leave vibrating matter and follow other paths, is somewhat analogous to the familiar fact of the conduction of heat—that is, of radiant heat and working heat, in causing expansion of substance. Thus the material substance of a bar of iron in becoming hot shows an expanding motion; and if touched with the finger will cause an intense burning sensation. It imparts the sensation of heat, simultaneously with the expansion of the iron; but the sensation of burning may be received without touching the bar—the subtile agent may be concentrated from its radiant heat, or from the sunbeam's heat, passed through vacuums and cause a severe burn. The movement of the molecules of the bar of iron may be traced in expanding the bar; but it cannot be said that these movements are essential to cause the burning—they are simply simultaneous facts with the moving of heat, just as the vibratory movements of gas flames and vibrating air are simultaneous facts with the movement of the agent that imparts the sensation of sound.

A common illustration of the vibratory condition of the atmosphere is to place a glass vase over an automatic bell and exhaust the vase of its air. On causing the bell to ring in the exhausted receiver no sound is heard outside, but on slowly admitting the air within the vase, first faint, then louder sounds are heard, caused as the air acquires its usual condition the full sound is again heard. In connection with this experiment, an additional fact may be mentioned which may be termed *bottling up sound*. After a small portion of air is re-admitted within the receiver, and the bell sends out a faint sound, if hydrogen gas be introduced within the receiver it mingles with the air and prevents the sound being heard; the bell still vibrates, but no sound is heard outside; the mixture of air and hydrogen absorbs the sound just as ice water will absorb heat without the heat rendering the water apparently any warmer—the water will absorb heat, but until the last drop of ice is melted, the absorbed heat does not raise its temperature. If air be admitted within the vase alone around the vibrating bell, sound is emitted; or if hydrogen be admitted alone, the sound of the bell is heard outside; but when the mixture of air and hydrogen surrounds the bell, the sound from the bell is absorbed by the mixture—the combination of the two fails to propagate the agent from the vibrating bell; but if a stick of wood or other conductor of sound be extended through the mixture and glass, the sound will travel through the stick and be heard. The same fact of concealed sound, or the agent that imparts the sensation to the ear, exists in a wire carrying sound; the agent has folded within the conducting wire its message from sight and hearing; but like the images or impressions of material objects, which reach the organ of vision from every direction, are unseen in the atmosphere on the route until revealed by a mirror or lens. So of the sounds carried by an electric wire, or sounds absorbed by a mixture of these gases. Both the wire and the mixture present a condition of matter that absorbs the allotropic light that is manifest in sound; but on altering the condition of the wire, making it non-electric by disconnecting the current, or by changing the mixture of the gases, sounds which each condition of matter concealed may then be guided into conditions of matter that reveals them.

The vibratory theory of sound appears to be held by a great many under the impression that it is a connecting link between matter and mind, between hearing and sound; and in extending it to the vibrations of æther, between objects and vision, between phenomena and sensation, this link is conjectured to be still further extended; not that there has been any absolute dependence or connection between feeling and rates of motion discovered, but as the rates of material vibrations increase and their oscillations become more rapid than the human mind can conceive, their motion becomes mysterious and incomprehensible; and these imperceptible motions are conjectured as offering a sort of northwest passage between facts and apprehension; between objects and sensation, phenomena and comprehension. But in the sensation of hearing, when no sounds are produced, or of the sensation of seeing things when there are no objects visible, or of the sensation of feeling cold when in a warm room, in these and other erroneous sensations their development is due to internal, instead of external causes. To say of these erroneous sounds, that what is sound within us, is outside of us motion of air; or to say of images of objects that appear when none such exist, that what is vision within us is, outside of us vibrations or undulations of æther, is evidently wrong, for the ear may be shut off from the atmosphere, and the eye closed from light, yet these sounds continue, and visions spring up before the eyes.

Not long after I had taken the extra dose of quinine, of which I spoke, I was out hunting squirrels with a gentleman by the name of M———, and while standing only a few feet from him, and no person within hearing distance, he suddenly turned to me and said, "Did you hear William speak to me?" I did not hear a sound of any kind; everything around seemed unusually still, and I answered that I certainly had not heard William speak, and assured him that I could not see how it was possible. William was his brother, who had suddenly lost his life only about a week previously by an accident, and had been buried several days. He insisted that he could not be mistaken in the voice; that it certainly was that of his brother William, and that he spoke in the most natural tone, and said to him, "Are you ready?" The incident made quite an impression on him, and the next day he took his gun and went alone to the same part of the woods. After being there some little time, and while sitting on a log, at just about the same hour, he was asked the same question in precisely the same voice, and when he came back he insisted that it was his brother's voice. He had both expected and feared that he would hear the voice before he went to the wood, but when the dead man's voice was really heard it made a very strong impression. The next day he was asked the very same question, at about the same time of day, in his own shop, there being others present, but none heard the voice. He afterwards went to the woods, hoping to hear the voice again, and expecting to ask several questions, but the question was never repeated, nor did he hear the voice again; but he ceased his excessive swearing. Not long after hearing him relate the incident several times, I had occasion to run nearly half a mile as fast as I could. I then sat down to rest a few minutes, and on rising from my seat I heard, or imagined that I heard, a band of music quite near, playing a piece of music that sounded precisely as it did in the circus of which I have spoken. The sound of the cornet, clarionet, trombone, bass drum, were each as distictly reproduced, and as plainly heard as when I was in the circus tent; yet there was no band anywhere in the vicinity, or even scarcely a sound near me. The music died out in a few minutes and vanished from my hearing. I knew a lady who lost three small children, in succession, from diphtheria, and for several months afterwards she continued to hear their voices around the house and yard; she felt confident that the children came back to visit and play. I also knew a gentleman who was troubled with catarrh in his head, and at times he would hear sounds resembling the puff of an engine.

These illusions of the sense of hearing are not referred to here as anomalous or unusual; on the other hand, they are quite common. Every one has either heard these false sounds, or has heard of similar experiences from others. The Apostle Paul was startled into sudden conviction of his evil life and wrong doing by a similar voice, asking, "Paul, Paul, why persecutest thou me?" Martin Luther, as is recorded in his history, frequently saw the devil and had several tussels with him, some of which were wordy and some ending in a trial of personal strength—he once throwing his inkstand at the vanquished and retiring devil. Joan of Arc heard, as she believed, the voice of St. Michael, and other holy persons, urging her on in her work of patriotism and religion.

Such illusions are easily brought about by the use of drugs of various kinds, and it is not many hundred years ago that drugs were sold for the express purpose of bringing the living spirit of man in communion and communication with the spirits of his deceased friends, through what is now well understood as illusions, produced by the temporary but disordered state of the nerves of hearing. The Mohamedans and Asiatics still adhere to this mode of communion with spirits. They usually commence by fasting for seven days, then retire to a lonely place, where they then burn incense of aloes, benzoin, mastic, hemp and various gums that tend to intoxicate, read certain chapters from the Koran at least a thousand times, and under this exhausting and disturbing influence of the nervous system, they seldom fail in rousing and sending allotropic light, as nerve force, over and through the nerves, with the enrollments of spirits, devils, images, sounds, odors, flavors, pains, pleasures and all the illusions of disjointed fragments of facts, fancies or thoughts which the various senses have enrolled in memory. Opium eaters, hasheesh smokers, and even alcohol drinkers still continue to find pleasure, and eventually pain in similar modes of producing illusions. It is well known to the medical world that very large and repeated doses of morphine, quinine and other drugs bring on chronic disorders of the sense of hearing, so that a constant humming noise is heard, accompanied with all kinds of sounds, and that illusions of each of the senses can be produced in a similar way.

It is a significant, although not an essential fact, that illusions of sight—and often of the other senses—as well as sudden disorders of the nerves, as in epileptic fits, that such disorders are usually ushered in and preceded by sudden flashes of light appearing before the eyes and often permeating the entire body. This is commonly the case with those who fall into the trance state. Persons having St. Vitus Dance often see flashes of fire before or within themselves, sometimes only dim halos of light surround them, at other times, this light is described as of a fiery red color.

The early history of Mahomet shows that he was an epileptic, and his early visions were accompanied with halos of light, sometimes mild and sometimes flashing balls of fire. In his lonely walks near Mecca, the hills and stones were at times all aglow with these delusive lights and figures; the air full of voices of spirits; the earth and rocks greeted him with, " Hail to thee, O messenger of God!" Like most epileptics, if not all, his trance states were preceded by lights and luminous flashes both within and around him. This luminous character of disordered nerve force cannot be deemed essential to illusions or delusions of the senses, for the senses of hearing, smelling, feeling and taste, if I have correctly interpreted their character, are founded on the non-luminous rays of light, and only that of vision on the luminous. But these flashes and halos of light from disordered nerves serve to confirm the statement or hypothesis herein set forth : That facts are absolute, and that knowledge is an enrollment of these absolute facts on the canvas and messenger of light. For we can then understand that when this messenger with its enrollment of knowledge and force is disturbed and roused from its dormant state in nerve matter, that the messenger with its enrollments should sometimes appear as lights, halos, fire balls, images and spectres; and that what is roused from an irritated or disordered nerve of hearing should correspond with sounds that have been received and gathered—enrolled as sentient knowledge as force may be embodied in muscles.

In galvanic batteries we see oxygen separating from

water and uniting with zinc, and in this and similar interchanges of the material elements we trace a simultaneous fact of the propagation of electricity. And again, when a bar of iron or other metal shrinks in cooling, or when water congeals, we also trace a motion of their molecules, and simultaneous with this motion a propagation of heat. We have in these facts a simultaneous motion of substance and the propagation of the subtile imponderable agents, and if I have correctly interpreted the phenomena of sound we have a corresponding fact of the simultaneous movement of a subtile agent propagated by vibrating air and other matter. This subtle agent that carries sound and the knowledge and information derived from sounds, as already put forth, is a non-luminous condition of light. And when we consider that sounds are taken up and continued by gas flames, also eliminated by what to the human eye are non-luminous rays emitted from a hot iron on a bar of lead, and that these sounds are taken up and carried by electric currents precisely as an electric current takes up heat, and magnetism and other subtile agents without an appreciable vibration, the inference that sound belongs to the group of subtile imponderable agents with light and muscular force is almost irresistible. We know that heat and electricity, magnetism and light are propagated by both the mechanical and chemical crash of atoms, and in the present state of knowledge we cannot say that sound outside of consciousness does not exist—but we can say that sound is propagated by vibrations. There is no question but that music, talk, all kinds of sounds are propagated through the air and elastic bodies by vibrations of their substance. These vibrations are an actual fact; they may be rendered visible, may be felt, and their relative numbers counted or ascertained; but in addition to their actual vibrations of material substance there is an invisible messenger of information to produce the sense of sound. In the phenomena of this sense, and also in the phenomena of each of the other senses, there is the working of the same broad principle which is presented in the phenomena of vision. We ignite a gas, or light a lamp and light is emitted. There is in the phenomena a clash of atoms. Oxygen and carbon meet, and we are able to trace an actual motion of their substance; but in addition to the clash of atoms of their actual substance there is also propagated the subtile illuminating agent that passes through transparent solids—an agent that is the actual enrolling agent and messenger of information to reach perception in the sense of sight. And so of the sense of hearing. In addition to the ascertained vibrations that propagate sound, there is not only the movement of material substance, but there is also a subtile agent—an agent that is the actual messenger to carry information and knowledge to the sense of hearing—an agent that consists of a non-luminous condition of light.

The explanation of delusive sounds, as well as delusive lights, developed by internal derangements of nerve matter, involves the discussion of polarized light. What can be said here is, that both luminous and non-luminous parts of the sunbeam becomes divided or polarized by the nerves of the senses; and that part of the beam that has gathered information carries its enrolled knowledge into the sentient nerves, while that part of the beam that embodies actual rearranging force becomes absorbed and retained in motor nerves—each part of the polarized beam being essential to the normal performance of a function. When, however, a nerve which holds re-arranging force becomes disturbed— developing only force without its sentient guide—tremors, twitchings, spasms, fits, &c., &c.,—as abnormal actions of the muscles occur; while if the sentient nerves holding information and knowledge becomes roused in abnormal process, the derangement startles the sleeping messenger into phrenzied displays of its messages of enrolled facts, thoughts and panoramic phenomena.

Smell, Feeling and Taste.

The discussion of the sense of smell, feeling and taste is here omitted as the principle already developed in discussing sight and hearing, is easily applied to each of the other senses, it being understood that odors and smell are founded on certain conditions of the same subtile agent having different modes of propagation. We know that odors are propagated by convection—conveyed on actual substance. As an illustration of the conveyance or convection of a subtile agent, we may heat one end of an iron rod red-hot and carry the hot end into gunpowder, and in this movement we perceive that the subtile agent of heat has been carried to the explosive. Or we may heat water until it becomes elastic steam; by the absorption of heat it acquires convecting power—the elastic power of the water carries the absorbed heat and delivers it to other objects of inferior temperature. So of odors, sufficient heat is absorbed and retained to render their substance volatile, and in conjunction with this absorbed heat their substance has absorbed non-luminous light that is carried and delivered instead of heat—the subtile agent having gathered information of those qualities that we recognize as odors, and that impart the sensation to the sense of smell by this messenger. An odor once smelt—as of a rose or an onion—is somehow retained in memory for reference and comparison, and what has been said of sight and hearing applies to this sense—a messenger has gathered information of properties and qualities of objects and enrolled them on its canvas and the enrolling agent, with its information imparted by odors, becomes stored as actual knowledge.

In one phase of the sense of feeling we perceive, and are conscious, of the transfer of the subtile agent of *heat*. But in addition to the capacity of this sense of feeling to detect heat or warmth, it also detects and reveals the forms and position of objects. In conjunction with the transfer of the agent of heat—an agent that reveals temperature—non-luminous light is also propagated, that carries and reveals information of other qualities of the forms and substance felt.

The sense of taste gathers information of other qualities of objects—qualities that are in the dormant inactive state, but that become active in the processes of solution and disintegration. Qualities of sweetness, sourness, acrid or bitter, &c., &c., exist in certain substances, and we know of their existence by the information propagated to this sense by processes that send the subtile agent with its messages from the dormant state into the perceptive nerves of the mouth.

In three of the senses, we have or can trace the action or manifestation of the subtile messenger of light. We have followed it into the apparatus of the eye, penetrating its actual substance a considerable distance. Then in the sense of feeling we find heat—one of the members of nerve force—actually traveling to impart information of warmth, and at the same time there is imparted information of other qualities of form, position, &c., and it requires no great stretch of imagination to suppose that these qualities are carried and revealed by the other members of the sun's beam.

In examining the phenomena of sounds and the sense of hearing, we have traced the development and propagation of sound in non-luminous light. Then if we attempt an analysis of the sense of taste, as when the tongue is touched with pepper, horseradish or wild turnip, we are startled by a thrill that traverses a large circle of nerves, and a part of this current thrill is, by presumption, due to the rearranging force contained in sunlight and condensed in the substance of these vegetables.

As the dormant state of knowledge and force has already been referred to a number of times, and will frequently be alluded to again, a few words here on dormant conditions will place the character of absorbed and dormant forces before the reader and give broader views of allotropic relations of force and substance; for whether we follow instinct, intelligence, or the interchanging capabilities of nerve force, we shall find each of the members of this trinity group from the sunbeam in the inactive and dormant condition, as well as in the active condition. We shall find information and judgment, memory, muscular force and vital processes, sometimes inactive, and with their presiding nerve force asleep.

Heat, as one part of the sunbeam, is frequently traced into inactive conditions remaining for indefinite periods without loss of energy.

ALLOTROPIC STATES OF ORGANIC FORCES.

Before passing to the direct origin or development of nerve force within the animal system, it is desirable that we shall have a clear understanding of the indirect origin of the force. As already said, this indirect origin is the sunbeam's rays. We can trace the absolute absorption of these rays into the substance of matter; follow in succession the change produced in matter by this absorption of the sunbeam's forces—this compounding of material and immaterial—then follow these compounds of material and immaterial into the forms of animate life, then trace the development of nerve force transferred from these compounds of material and immaterial, these unions of matter and the sunbeam's rays, and again find that the material substance, after transferring its absorbed agents and forces to the nerves of animal life, is again in its previous condition of inorganic refuse, again incapable of furnishing vital forces to organic life; but again susceptible of again absorbing the sunbeam's agents and forces.

This non-ending circle of absorbing immaterial light by material substance produces one of those conditions of matter which are here termed allotropic, while the transfer and discharge of the absorbed forces and agents produces another but different condition of the substance, also called allotropic—the condition of the substance in which it holds the sunbeam's rays, adapting the substance for organic life—the other condition of the substance in which these absorbed agents are transferred and discharged, rendering the substance unsuitable for organic, but adapting the substance for inorganic and inanimate compounds.

This view of the character of allotropic conditions of matter is, as far as I am aware, now presented for the first time. It is however well known to chemists that the primary elements of matter have what have been termed allotropic conditions, as of hardness, fluidity, flexibility, &c., &c. These facts of allotropic states of matter are not novel, but what is insisted on here is, that these allotropic states are absolutely and positively due to the absorption and retention of immaterial agents, and the compounding the material with the immaterial, the union of matter and force, the combination of agents and substance. Neither is the fact that immaterial agents can be and are absorbed by material substance novel, or a new discovery; but what is insisted on here is that by the absorption, transfer and discharge of the sunbeam's group of forces by and from substance, the agents and forces undergo a corresponding change which I have termed allotropic change and conditions of the agents and forces.

The stray facts of allotropic conditions of matter, and the disconnected facts relating to allotropic conditions of force are quite familiar to students; but their mutual relations, and their capacity to furnish a full and clear explanation of the dividing line between organic and inorganic nature, between the animate and the inanimate, between the living and the dead, are herein presented as novel. As the indirect origin of nerve force, the sunbeam's group of forces are absorbed by material substance, remain dormant for indefinite periods of time, then by allotropic changes these absorbed rays reappear as nerve force—their development as agents traversing animal nerves, being due to the allotropic changes of the material substance which have absorbed and retained the sunbeam's rays.

In this discussion of the identity of light and nerve force, these conditions and changes of matter, and these transfers of imponderable agents and forces, which I have termed allotropic, have been frequently referred to for the reason *that allotropic conditions and allotropic changes* give us the foundation for our real and positive knowledge of the real dividing line between the organic and the inorganic, the living and the dead. These allotropic conditions and changes not only furnish facts for clearly drawing the dividing line between the animate and the inanimate, but they also enable us to follow in succession the rearranging forces of organic life in the vegetable kingdom; then the combination of rearranging force with locomotive force in low orders of animal life, followed in higher orders with insticts; then with grades of intelligence, finally followed in man with miniature sparks of omniscience.

The doctrine of organic forces, as taught by many leaders, is thus stated by Professor Tyndall: "The building up of the vegetable then is effected by the sun, through the reduction of chemical compounds. The phenomena of animal life are more or less complicated reversals of these processes of reduction. We eat the vegetable, and we breathe the oxygen of the air; and in our bodies the oxygen which had been lifted from the carbon and hydrogen by the action of the sun again falls towards them, producing animal heat, and developing animal forms."

" The nature of the animal body is that of organic nature. There is no substance in the animal tissues which is not previously derived from the rocks, the water and the air. Are the forces of organic matter then different in kind from those of inorganic matter? The philosophy of the present day negatives the question. It is the compounding in the organic world of forces belonging equally to the inorganic that constitutes the mystery and the miracle of vitality. Every portion of every animal body may be reduced to purely inorganic matter. A perfect reversal of this process of reduction will carry us from the inorganic to the organic; and such a reversal is at least conceivable." " The tendency of modern science is to break down the wall of partition between organic and inorganic, and to reduce both to the operation of forces which are the same in kind but which are variously compounded."—*Tyndall in Fragments of Science.*

It is always a pleasure to read or quote from Professor Tyndall on the subjects of which he treats whether we differ with his views or not; for whether it be facts or philosophy, his position is stated fairly and clear. In the above quotation there is no ambiguity, no uncertain meaning to the words, and it is undoubtedly true that for a considerable time many prominent writers have taught that there was no real dividing line between the animate and the inanimate; but with these views and the inference which I have drawn from allotropic facts, there is direct conflict; and in a conflict for supremacy one fact is worth a bushel of theories. So far from a proper grouping of modern facts tending to break down the wall of partition between the organic and inorganic worlds, both the facts and all legitimate inferences from them tend to build and establish a still more impassable barrier-wall between the organic and inorganic, between the living and the dead. The inference so frequently drawn that because the human body is composed of substance found in rocks, and water and air, it can contain nothing but what rocks, water and air contain, is completely overthrown by the facts of allotropic relations. The allotropic conditions of matter and the allotropic conditions of force enable us to trace clearly and distinctly the distinction between organic and inorganic matter; they enable us to understand clearly how the same material elements can be at one time endowed with life, and at another time totally devoid of life; at one time endowed with heat, at another time totally devoid of heat; at one time endowed or charged with magnetism, then with the magnetism totally discharged; at one time endowed or charged with chemical affinity and chemical properties, at other times totally devoid of chemical affinity and chemical properties. In other words, the allotropic conditions and states of matter show an absorption, and a retention of imponderable agents, and receive therefrom certain properties, and at other times by the discharge of these agents and absorption of other agents their properties become changed; properties imparted by one agent vanish by the discharge of the agent to be superceded by other properties imparted by other agents—the organic forces of life, muscular force, nerve force, viability, &c., being in this respect like magnetism, electricity, chemical affinity or heat, capable of being transferred from one substance to another, and remain with greater or less degrees of permanency.

If we hold a common sun-glass so as to concentrate the rays of the sun on the hand, we soon experience a sharp burning sensation; heat concentrated in this way has been made to cook meat and to run engines. If the heat be concentrated on a lump of ice the heat is absorbed by the substance of the ice, and, although large amounts of heat may become absorbed by the ice, it does not of necessity heat the substance the slightest perceptible degree—it simply alters its structure—it changes a hard solid to a mobile liquid.

Allotropic States of Organic Forces—Allotropic Conditions of Heat—Latent Heat.

The inactive condition of heat was first pointed out by Dr. Black; he noticed that heat was absorbed and retained by ice-water without affecting the thermometer. In Dr. Black's original discovery, he found that by mixing hot water with an equal quantity of water at ordinary temperatures, the mixture would have an intermediate temperature—the hot water would lose just as much heat as the cool water would gain, but if the hot water was mixed with an equal quantity of frozen water the mixture would not show an intermediate temperature; there would be no rise of temperature of the mixture until the ice was all melted; the ice appeared to absorb an amount of heat and pass it into an inactive, dormant, insensible state or condition. This fact is stated in Miller's Elements of Chemistry in a very clear manner, as follows:

"Disappearance of heat during liquifaction, when matter passes from the solid into the liquid state, or from the liquid into the aeriform state, heat in large quantities disappear, and ceases for the time to affect the temperature; hence this modification of heat is called latent heat. For example, when a lump of ice at thirty-two is brought into a warm room it gradually thaws and is converted into water; but neither the ice nor the water in contact with it rises in temperature; so long as any portion of the ice remains unmelted the water continues to indicate the temperature of thirty-two, as does also the ice. Again, a pound of water at 212, mixed with a pound of water at 32, gives two pounds of water at 122, which is the mean temperature; but a pound of ice at 32, mixed with a pound of water at 212, gives two pounds of water, of which the temperature is only 51. In this case the water has lost 161 degrees, whilst the ice has gained only 19 degrees, so that 142 degrees have disappeared, or have become latent. Hence in order to convert a pound of ice at 32, into water at 32, heat sufficient to raise 142 pounds of water from 32 to 33 is needed. This heat, however, is not lost, for if the progressive cooling of water be observed in an atmosphere many degrees below the freezing point, it will be found that the temperature of the liquid sinks regularly until it reaches 32, when it becomes stationary and freezing begins; the heat being supplied by that which is latent in the water. As soon as the whole has become solid the thermometer again shows that the temperature of the mass sinks until at length it reaches that of the surrounding air; some idea of the quantity of heat that is required to convert ice into water without any apparent rise in temperature may be formed from the fact that the simple conversion of a cube of ice three feet in the side into water also at 32, would absorb the whole heat emitted during the combustion of a bushel of coal."

The latent heat of water is greater than any other known substance, but all other substances are known to acquire latent heat; that is, heat will be absorbed by them and disappear, and remain inactive for a time and afterwards become free and produce heating effects. The source of heat for this purpose is immaterial; we may concentrate the sun's rays on a cake of ice and they will be absorbed without the ice becoming sensibly hotter. When crystalline salts dissolve in water heat becomes latent and disappears; advantage of this fact is taken in making freezing mixtures; one of the most common and convenient being ice or snow and salt, the salt in dissolving absorbing heat from any body in contact with it. There are a variety of circumstances that influence the conversion of latent into free and sensible heat; if we expose a hot saturated solution of salt, made by mixing snow with the salt, to the vapor of boiling water, the steam or vapor condenses in the solution, and although the solution was only 212 and the steam was only 212, the solution will become hotter and rise to 250 by the latent heat of the solution becoming free and sensible; or we may make a saturated solution of Glauber salts hot and pour it into a bottle, cork it tight, and let it cool; after a time on removing the cork the salt will suddenly crystallize, the latent heat becoming free and making the mass quite warm.

We can form a fair conception of the difference between active and inactive heat by touching the finger to a bit of coal no larger than a pin's head and contrasting the effect on the finger with the effect produced by dipping the finger into a glass of water; the glass of water may contain a very much greater amount of heat than the bit of coal, but its immediate effects are imperceptible, while an intense burning will be derived from the coal. Every one is familiar with the large amount of heat evolved by the burning of a pail of coal in a common stove; yet all the heat evolved by this amount of burning coal is readily absorbed by a tub of ice without raising its temperature a perceptible degree. The most delicate thermometer will not detect the slightest increase of temperature from the heat absorbed and passed into the latent state in ice water; a thirsty camel would readily drink the absorbed heat obtained from a pail of burnt coal and manifest a feeling of comfort and satisfaction after swallowing this combination of material substance and immaterial force.

Our animal heat is derived from what we eat, drink and breathe; it exists in these substances by reason of having been absorbed from sunbeams, and is developed from the latent state by the processes that take place within the body. In these processes heat is more especially developed from food than from water—water serving rather as a regulator than as a source of heat; rise of temperature being prevented by evaporation from the skin.

Allotropic States of Organic Forces—Agents Other Than Heat.

What has been herein presented of the imponderable agent of heat is also true of each and all imponderable agents, including light with its trinity group of illuminating, heating and rearranging force. It is well known that light is absorbed by vegetation, producing perceptible effects in changing its color and vigor of growth. When a board is laid on growing grass so as to exclude light, the grass soon presents a bleached and sickly look, that is rapidly changed to a vigorous and healthy appearance by removing the board and allowing the grass to absorb light.

By the absorption of heat by ice, we perceive that the absorbed heat makes a perceptible change in the substance—converting a hard solid to a mobile liquid. And again in the absorption of light by vegetation we perceive that a change is effected—changing its color and imparting vigor. In these processes, by which imponderable agents are absorbed, we perceive allotropic changes of substance due to the absorption and detention of imponderable agents in dormant states—for neither the heat in the ice water nor the absorbed light in the vegetable are appreciable to immediate perception; but the absorbed agents have produced perceptible effects—allotropic changes of substance; and in the processes that develop these dormant agents and bring them into the active states again, we can trace allotropic changes of force—changes from the dormant to the active conditions. Around this central fact of allotropic change are grouped organic structure and organic life.

It is pretty generally known that in connection with the luminous rays of light there are non-luminous rays—rays that do not excite the organs of vision. Some of the properties of these invisible rays have become familiar to the public in the art of photography—photographic impressions being largely due to invisible rays.

Tyndall, in Scientific Materialism, says: " Two-thirds of the rays emitted by the sun fail to arouse the sense of vision. The rays exist, but the visual organ requisite for their transmutation into light does not exist, and so from this region of darkness and mystery, which surrounds his rays may now be darting that which requires but the development of the proper intellectual organs to translate them into knowledge, as far surpassing ours as ours surpasses that of the wallowing reptiles which once held possession of this planet. Meanwhile the mystery is not without its uses; it certainly may be made a power of the human soul, but it is a power which has feeling not knowledge for its base, it may be, will be, and I hope is turned to account both in steadying and strengthening the intellect and rescuing man from that littleness to which in the struggle for existence or for precedence in the world he is continually prone."

Some of these invisible rays, as already noticed, determines an allotropic change in the properties of matter, changing for example, carbonic acid from a substance charged with chemical properties and fitting it for the uses of organic life through the leaves of vegetation. It is through the influence of the invisible rays of heat that ice is changed to water.

Instead of adopting the view presented by Tyndall in the above extract, of invisible rays requiring the development of other intellectual organs for their translation into knowledge, it is here suggested that they are the source of instincts. The force that becomes manifest in animal nerves is so completely interwoven in organic structures that its study involves the entire plan of organic life. The nerve force not only traverses the nerves of animals, but it actually permeates the entire forms of organic life. The tiniest moss or mould that overspreads decaying forms of life, holds within its material molecules a share of the *force* and *sentient* group. These low forms, in their instinctive modes of carrying out their plan of life, reveal incipient intelligence and executive force. These qualities of living things are no more inherent properties of the substance of which they are composed than fluidity is the inherent property of the substance of which water is composed. In order that the substance of which water is composed shall become fluid it must absorb and retain a fixed and definite amount of absolute heat; and in order that this substance shall assume the gaseous state of steam, it must absorb and retain another fixed and definite quantity of heat. The retention and fixation of these definite amounts of actual heat, first in forming water, then a larger amount in forming steam produces the allotropic states of fluid, and gaseous oxide of hydrogen. In extending this principle to the actual absorption and fixation of light and other agents in material substance, it is here claimed that the substance envelops and retains the agent in dormant inactive states for its future appearance as nerve force.

As already said, the sentient and force qualities manifest in nerve phenomena are no more confined to nerves and nerve substance than electrical phenomena is confined to connecting wires. It is well known that the subtile agent of electricity that becomes manifest in the metal connections of batteries is developed and sustained by certain conditions of matter that must be maintained for its effectual development. And so of the subtile agent that traverses the nerves of animal life—the nerve current is developed and maintained by certain interchanges of material substances that hold condensed light in a dormant state. But in addition to the development of electricity by galvanic batteries, and its manifestation on connections, it is also manifest within the battery, and can also be developed by friction, and by other modes. And so of nerve force, it is manifest in other conditions of matter besides nerve matter. The three prominent qualities of nerve force are presented in the trinity group of apprehension, knowledge and force; and we can find manifestations of this group, and each of its members, in organic matter devoid of nerves. In vegetation there is manifest a sort of incipient intelligence termed instinct. We might define instincts as tendencies of organic molecules and structures from internal forces to carry out the plan of organic life. This plan involves both knowledge and power—these qualities are grafted on to material substance by means of the agent in sunbeams.

In vegetation and the low forms of organic life, forms which have no special organs for transmitting nerve force, a sort of incipient intelligence is manifest, but the mode of manifestation is different from the mode that obtains in animal life. In sprouting seeds we can trace a sort of incipient intelligence, directing the root downward, and the stem and leaves upward. Additions to this incipient intelligence may be traced through all the stages of its after life. It selects its food by this engrafted intelligence, showing periodical desire and satisfaction when gratified. Placed in a dark cellar, the growing stems of potatoes and common vegetables grow towards the light, taking the shortest, nearest direction to reach it—and it is a well ascertained fact that the light thus sought for, by growing towards it is actually absorbed. Roots of vegetation, especially those fond of watery soils, will make quite short and abrupt turns to reach a springy soil; this instinctive growth of roots, and stems, and leaves, and flowers for food, and water, and light is not less certain than the instinct of ducks for water or of a horse for oats, being well known and familiar, but the mode in which this instinctive intelligence produces its manifestations is different from the mode in which intelligence is manifest by the animal. Visions of water reach the optic nerve of the duck, and he is provided with a muscular apparatus enabling him to reach the watery luxury. The vegetable has no optic nerve to receive impressions of the distant water, yet that it is endowed with a percipient faculty for apprehending its presence beyond its roots is certain; but its mode of reaching the objects of its desire is through the slow process of growth with vegetation.

These vegetable instincts are referred to here not as being new or novel, but to fix attention to the important fact that instinctive intelligence, instead of being an inherent property of carbon or other elements, is grafted on to allotropic states of these elements by means of *condensed sunbeams*—for until the material elements have absorbed sunbeams they do not manifest instinct or intelligence; and it should be noticed in this connection that material elements also are not endowed with inherent *chemical properties or affinities*—such affinities only being grafted on certain allotropic conditions of the elements, and may be displaced and superceded by other affinities and properties. By keeping in view the fact that as a medium of intelligence, light is not destroyed by impinging on the earth or other non-transparent bodies, but is absorbed and still exists in an allotropic form, and still capable of transmitting intelligence of distant objects to organic life, these phenomena of roots growing towards distant food, of flowers opening to distant light, or of tendrils reaching out to grasp supporting stalks become less mysterious. The same agent that delineates on the optic nerve of a duck, visions of watery luxury, is the same agent that permeates the soil and reveals to the instinctive apprehension of the roots of plants the food and watery object of its desire. Light diffused through the soil with heat or warmth forms an unbroken line, over and through which information travels between distant substance and inner desire. Light that is absorbed by the earth passes into the latent state just as heat of the sun's rays pass into the latent state when absorbed in the earth's substance. When this latent heat that exists in the soil is rendered free, the freed heat becomes a source of warmth to surrounding objects by the process of diffusion; and so of the force of light that lies latent in the soil when it is thrown into the active state, it still retains its property of transmitting intelligence, and becomes a medium to these low instincts, and carries information of qualities adapted to their wants, cravings, and desires. It is still the messenger of knowledge; it enrolls knowledge on its canvas, and in all cases it is the imponderable agent that transmits and holds information, and not the material that envelops it that transmits knowledge. The substance of nerves enables the agent to move in its allotropic state by conduction; but when thus moving through the nerve substance, the messages still cling to the messenger—they still adhere to the agent. In the broad fact, already noticed, that all imponderable agents have five distinct modes of transmission or propagation, we perceive that other substance than nerve substance transmits or propagates nerve force. In its original mode of radiating from the sun, light is propagated two hundred thousand miles a second, but in its latent state it is motionless. Then in another condition, as propagated with warmth through vegetation, it moves less than two feet an hour, while in being propagated through animal nerves its rate of motion is about one hundred feet a second. In vegetable life warmth is distributed by convection; it is carried on the circulating fluids, and this is the principal mode in which the forces of vegetable life, including allotropic light, move. Information and knowledge of the presence of nutriment is diffused through the soil on its messenger, just as warmth is diffused through the soil, and reaches the circulating fluids of plants and rides the nutriment molecules with the same motion which these molecules have in the process of growth, directing them towards the objects of desire. Knowledge that would be conducted through the nerves of animals quick as thought, is only transmitted through the vegetable by convection; that is, it is conveyed on the molecules for growth, and moves by and with their motions.

In conjunction with the qualities of light to enroll information, there are properties that embody chemical force—rays that embody a power for rearranging the elements of material compounds—rays that carry a force to produce allotropic changes of the elements—a force that will discharge inorganic affinities and substitute organic affinities; a force that, in the vegetable kingdom, determines the absorption of carbonic acid in the leaves, and in the animal determines its expulsion. This chemical, or rearranging force, contained in sunbeams with heat and illuminating rays, is non-luminous, and in organic processes is an *allotropic condition of chemic force.*

These invisible chemic rays of light, that produce such a variety of phenomena among material molecules and primary elements, have several names; they are sometimes termed chemic rays, actinic rays, photoic, rearranging, allotropic, &c., &c. By their absorption into material substance, they become the executive agents of function, and processes of organic structure; they are the internal forces that determines the affinities of primary elements, and the instinctive tendencies of organic molecules. Under the guide of inherited intelligence the executive rays produce organic forms and instinctive actions. In carrying out the phenomena of organic life they are essential members of the group of forces; but it is not to be understood that the three qualities of the sunbeam, to which attention is being drawn, constitute the whole of what is known as organic life; there are other members equally important conjoined with nerve force to carry out the sum total of vitality, or life. Oxide of hydrogen, after it has absorbed heat, becomes adapted to organic processes; but until it has actually absorbed heat sufficient to render it fluid, it is a hard crystalline solid, and as useless to organic life as paving stones. It is by virtue of its absorbed heat, as an imponderable agent contained in its substance, that water becomes so essential to organic life. The same fact obtains with the allotropic rays; they alter the nature of elements by imparting essential qualities—imparting organic affinity, and rearranging force. This class of rays forms a connecting link between organic and inorganic matter. The property of inducing chemic changes, which the actinic rays of light exert on plant matter, is well ascertained and known as positive facts. The influence of these rays is readily traced in altering the allotropic states of all the primary elements that enter into the structure of organic forms and of altering the structure of their combinations. The elements of carbon, hydrogen, oxygen, nitrogen, sulphur and phosphorus are all known in different allotropic states conferred by the absorption of light. Then, in addition to this first essential influence which the invisible rays exert in altering the allotropic states of the substance, there is conjoined with the alteration of structure an imparted susceptibility to the illuminating rays. There is first an absorption of the invisible rays that alter atomic structure and retain in the substance these invisible rays, then, by virtue of properties imparted to the substance by these absorbed rays, the substance responds to the illuminating rays. The apprehension, or prehension, by plants without the assistance of special organs of sense—the apprehension of the presence, at a distance, of food and drink is well established by common observation. The blanched leaves and vines of plants that have grown in the dark possess no visual organs for distinguishing illuminating from non-illuminating rays, yet potato stalks, twigs, grass and all kinds of growing leaves of vegetation, will make quite appreciable efforts to reach the illuminating rays by bending towards them in the process of growth.

We can readily observe the plain and positive influence which light, as an imponderable agent, exerts on that exquisite organ of sense, the eye. The front part of the eye is covered and protected by a muscular, colored curtain termed the iris; it is that part of the eyeball that gives to different eyes their different colors. This muscular curtain has a small central opening termed the pupil, and this opening or pupil instinctively enlarges, in a dim light, to allow a greater amount of light with its messages to penetrate the ball. Then again, the pupil instinctively contracts in a bright light to shut off a portion of the light. These instinctive and spontaneous movements of the muscular fibres of the iris are brought about by the influence of light on the fibres of the curtain, and these instinctive adaptations of the pupil are essential to distinct vision. A similar movement of organic fibres, in response to light, is seen in the opening and closing of flowers by light and darkness. The fibres of the eye's curtain open for the reception of messages of information carried by one quality of light, the fibres of the flowers open for the reception of rearranging force carried and embodied in another quality of light; the eye receives sentient messages, the germs within the flower receives actual force; and both the enrolled knowledge and embodied force are absorbed and retained for future use in the current of animal nerves. In plants, substance that is sensitive to light, and substance that is unsensitive are united, forming horticultural elements. By reason of this combination, sensitive movements of plants in response to light is necessarily a movement of the joint combination of both sensitive and unsensitive parts, resulting in a movement by growth towards food and water. In the animal system these sensitive and unsensitive parts are separated, forming two or more classes of organs, as in the muscles and nerves. In the scale of organic life, we first trace functions performed by material combinations of a primary character—functions which in higher organizations are more intensely manifested and carried out by distinct organs. In some low classes of animal life, physiologists find circulation and digestion carried on within the same cavities or organs. Instead of there being a heart and stomach there is only one organ for the two processes of impulsion and digestion. In other classes of animals respiration and digestion take place in the same cavity or stomach.

The two-fold function of chemic rays, of altering composition, and, at the same time, inducing a rearranging change of substance at a distance, may be traced at the primary absorption of chemic rays by the leaves of plants; the rays not only alter the composition of the leaves by changing their colors, but the rays also impart an impulsive throb of vigor to the entire plant. The invisible rays exert an influence that reaches from the leaves through the entire fibrous structure to the roots; and what is remarkable in this connection is that the influence from each leaf follows its own line of fibres as accurately as nerve force follows its line of nerves.

The group of forces found in the connecting wires of galvanic batteries are collectively called a galvanic current. This group embraces heat, and light, and magnetism, as well as electricity—currents from different batteries holding different proportions of each of these agents. The subtile agent and influences traversing animal nerves called nerve force, embraces all the subtile agents contained in a sunbeam. The allotropic conditions and changes of these subtile agents may be followed and traced in their successive changes in a very simple manner, showing analogous properties in each condition. If we stand in a perfectly dark room containing people and different objects, we see nothing in the darkness; but if we then light a lamp, which may be held in the hands, the light immediately diffuses through the room and reveals to the eye the people and all the different objects in the room. Discarding all theories of what light is, and examine only its properties by what it accomplishes, we perceive that the light emitted from the flame of the lamp has the property of illuminating different objects and rendering them visible; we perceive that the light passes to the objects and is then reflected back again loaded with information of facts and phenomena. The light emitted from the lamp after traversing large spaces comes back with information of what is to be seen, and reveals these facts to the eye.

This property of traveling through surrounding space and gathering information of the panorama of facts is but one of well known properties of light; but this property is a marvelous one. A superior being placed in the central sun could, by the aid of this property, instantly perceive what is taking place on the earth, moon and stars. The rotation of the planets brings into view in succession the entire surface of each; by the assistance of the property of refraction, diffraction, and reflection, the light emitted from the sun returns with information to the central sun, substantially as light from the lamp illuminates and reveals different objects. This power of carrying information is one of its properties familiar in the phenomena of nerves. If a fly, or other irritant, injures the foot or hand, information of the fact is sent through the nerves from these parts to nerve centres, where it is determined what shall be done with the irritating object. But this property of enrolling information of fact and carrying it through space or through the animal system is but one of its marvelous properties.

In addition to this property light embodies actual force —it embodies properties that gives it executive ability; it embodies the property and power of heat; and it embodies the property of rearranging the molecules and combinations of material substances; it embodies a property that has the power of changing the character of all known substance; it slowly alters the character of even the hardest glass, as well as the delicate bloom on fruit, and of a child's cheek. This property is sometimes known as chemic affinity, sometimes as actinism, and sometimes as the rearranging force of sunlight. In some cases it induces chemical combination; in others it induces decomposition; in others it induces allotropic changes of material elements. These several phenomena of chemic affinity, actinism, organic and inorganic interchanges, are substantially allotropic changes, founded on the property of light to produce allotropic changes of rearrangement. If we invert a glass vase over a common looking-glass on a table before a window, arranged so that a beam of light from the sun shall impinge on the vase, we perceive that the glass of the vase does not stop the beam of light, but it passes through the glass of the vase and impinges on the mirror and through the glass of the mirror, and on reaching the metalic back of the mirror its course is altered, and it is reflected in other directions, returning through the mirror and vase glass. If now we hold an apple, a bunch of grapes, a picture or stand before the glass, the light that illuminates these, or other objects, gathers information of their several qualities and carries the information through the glasses to the metal back of the mirror, when both the light and its information is reflected out through the vase precisely as the light of the sunbeam was reflected out. Both the light and its enrolled pictures pass readily through the glass to the eye. This property of light to enroll and carry information is well known, and is associated with the power of heat and actinism.

If we substitute for the mirror a glass vessel containing ice, we can trace the fact that the sunbeam not only enrolls information, but that it also embodies heat as an actual force. Letting the sunbeam through the window and impinge, as before, on the vase, it passes through the glasses to the ice. The beam, on meeting the ice, parts with heat, a certain amount of which is absorbed—the absorbed heat converting the ice into water; and it is well to notice here that a quart of ice will absorb a certain amount of heat, but that two quarts will absorb twice as much. Now it is a well known fact that the heat that was absorbed by the ice exists and actually is retained in the melted ice water; yet a finger placed in the water will find it just as cold as the ice, and a delicate thermometer will not detect the slightest difference in temperature between the ice and ice water. That the heat actually exists as an actual force in the water is well known and is easily proved; for if we surround the ice water, or place a vessel above and another below, containing ice and salt or other substance colder than ice water. the heat that was absorbed and retained by the water will be expelled and radiated into surrounding space, where it can be concentrated by concave mirrors and made to explode powder, melt metals, or run steam engines.

This absorption of heat has been adverted to several times for the simple reason that what happens to this member of the sunbeam is just what happens also to each of the other members of the group—they are each and all absorbed and retained in the dormant inactive state by vegetation, to reappear as nerve force. If we substitute for the bottle of ice under the vase, a bottle containing a mixture of chlorine and hydrogen gases that have been kept in the dark, and let the sunbeam impinge on the mixture, chemical force is absorbed by the mixture and the two gases combine, and form a single compound—chloride of hydrogen—a compound that will absorb a further amount of chemical force substantially as ice water absorbs and retains heat. A still better compound for absorbing chemical force from the sunbeam, than chloride of hydrogen, is chloride of silver. This compound, by its absorption of chemic force becomes changed from a white color to a reddish brown, enabling the eye to follow its absortion.

The character of chemical or actinic force absorbed by chloride of silver and similar compounds has been noticed in discussing muscular force, and is here omitted; but its property of effecting decomposition is readily seen by substituting nitrate of silver, in solution, for the bottle of chloride; when the sunbeam impinges on this solution the silver salt is decomposed, and metallic silver precipitated. The double nature of actinic rays of light may readily be seen by moistening a clean slip of ivory with a solution of nitrate of silver and exposing it to the sunbeam. Chemic force is absorbed and the ivory blackens after a few hours exposure to the light, but on rubbing the blackened surface, a coating of pure silver will be found deposited on the ivory.

The relation of light to nerve force, however, can be traced in a more direct line through vegetation than through inorganic matter. If we place a plant under the vase, letting the light impinge on its leaves and flowers, each of these members of its group are absorbed and produce visible effects; the struggle of blanched stems and vines to reach the luminous part of the ray, the opening of flowers, and the absorption of the entire beam in seeds, are familiar facts. The force of actual heat, and the rearranging force of organic affinity, thus absorbed by vegetation, are retained in the horticultural elements, are carried in the material substance into the animal as food, and reappear in the group of forces that traverse the nerves of animal life.

The absorption of heat by vegetation and its reappearance in the various processes and functions of animal life is a fact somewhat more familiar to students than the absorption and reappearance of the luminous and rearranging members of the sunbeam's group, but is no more certain as a positive fact. The dormant, inactive state of heat in vegetation, enabling it to be carried into the animal, is an allotropic condition of inaction which each of the other members assume in the processes of organic life, that have been partially noticed, and will be noticed more fully in another place. The fact to which attention is here drawn is that these several agents in becoming absorbed in vegetation simply pass into an allotropic condition of inaction, and are not transmuted into potential, nor any other energy.

By drying, or depriving the vegetable matter of water, the absorbed heat, and the absorbed light, and the absorbed allotropic altering rays may be readily developed or set free by combustion. The absorbed altering force becomes manifest at proper temperatures by inducing combination with oxygen—the combination developing heat, which becomes manifest by expanding the gases formed by combustion, and the light becomes manifest by illuminating surrounding objects. This mode of developing the absorbed agents, however, is abrupt and simply shows that the agents have been absorbed and stored in the vegetable substance. In the processes of animal life these abrupt chemical changes do not occur; their development in the animal body is more analogous to the development of light and heat and chemical force within galvanic batteries, in which the developed forces become grouped in the form of and are propagated as current instead of being propagated by radiation.

We can trace, in succession, the five allotropic states of heat, and by so doing can form a more concise view of the five allotropic states or conditions of light. If we suspend a metal kettle, filled with water and floating ice, over a fire we can observe that heat radiates from the burning coal, impinging against the bottom of the kettle. The velocity of the radiating heat is very great until it impinges on the metal; its motion is then checked and retarded, but not stopped; the impinging heat, instead of radiating through the metal passes through by the second mode—that is, by conduction; it moves or is propagated from atom to atom along the substance of the metal. In the first or radiant mode, the heat moved without the assistance of material substance; in this mode it will pass through a vacuum; but in the second mode it is conducted by the substance of the kettle to the water within the kettle, where it again changes its mode of motion. Impinging on the water, it finds an almost non-conductor and takes on a new condition or mode of propagation; it gathers on the molecules of the fluid and is propagated by the mode known as convection; it is neither radiated nor conducted, but both the molecules of water and the agent of heat rise up through the surrounding fluid in union. When this combination of heat and the molecules of water impinge on the floating ice within the kettle, the heat takes on another mode or condition. The molecules of rising water part with the heat which they were conveying to the ice in which they become latent. The heat thus rendered latent by absorption converts the solid into a liquid, and as long as the molecules remain liquid the absorbed heat will remain dormant and without apparant energy; on crystallizing the molecules of water, holding the absorbed and latent heat, it is expelled and is again changeable to each of its five modes without the slightest loss of energy; but there is no known mode of changing it into light or any other kind of energy; *it can neither be blotted out of existence nor transmuted into life, electricity, or any other imponderable agent.*

There is another state or condition of heat, which is here omitted, but each are allotropic conditions, as each are convertible into the other conditions. Each of the other members of the sunbeam have five analogous states, or conditions. The illuminating rays and the rearranging rays may be traced in the radiant condition, then as being conducted, then as being conveyed by convection, then as in the dormant or latent state, and *in the processes of organic life each mode becomes available, and each is used to alter the properties of, and to control material substance.* Instead of taking the radiant heat from a burning fire to impinge on the kettle of ice water, we may let the sun's rays impinge on the kettle, the impinging heat passing in succession through these several modes. Instead of following the heat rays we may follow the illuminating rays through these allotropic changes. The heat rays, on impinging on the metal, becomes absorbed and temporarily retained—absorbed and then emitted. A similar fact may be observed of absorption and emission of light from loaf sugar, diamonds, different kinds of spar, and even the waters of the ocean.

The fifth mode or state in which imponderable agents exist may be termed the phosphorescent state or condition. When light impinges on sugar, diamonds, spars, &c., it is absorbed, and in being emitted it passes into this phosphorescent mode. Heat may also be traced as passing into this mode, but this mode is more familiar in its relation to illuminating light.

In tracing the several modes in which the agent that traverses animal nerves exists this phosphorescent mode becomes important. Illuminating rays are absorbed by sugar, diamonds, and other bodies substantially as heat, impinging on a metal pail, is absorbed in heating and being conducted through the metal—the rays of heat heat the metal; and so of the light rays absorbed, they illuminate the absorbing substance. The evolution, or emission of the absorbed light from the sugar may be observed by placing the sugar, after it has absorbed the light, in a dark room, when the light will be slowly emitted. This slow emission of light resembles the slow emission of heat from a heated metal.

Light absorbed by the waters of the ocean is also slowly emitted at times. This is noticeable when the water is disturbed by ships or the paddle wheels of steamers. The crests of waves show a gleamy white phosphorescent light that illuminates the disturbed water ; and disturbed in this way from its dormant state in ocean water, it has the appearance of being roused from its sleep with reluctance, and, instead of dashing off in the radiant mode soon sinks back into the water with a sleeply stupor. A still better illustration of light in the dormant state may be seen by standing before a looking-glass in a dark, cold room, and crushing rock crystal candy between the teeth ; during the crushing process the mouth will be filled with a reddish colored light, sometimes glowing so fiercely as to give the mouth the appearance of being on fire. Similar changes of allotropic states of light from the latent to the phosphorescent state may be seen by saturating hot water with Glauber's salt and setting it aside to cool; the salt will remain dissolved until it is distrubed, but on dropping into the solution a few grains of sand, the salt will suddenly crystallize and evolve light during the crystallizing process—an effect that is entirely analogous to the evolution of heat by the crystallization of water and conversion into ice—the crystallizing process evolving the latent heat from the water. A similar evolution of absorbed latent light from frozen salt may be seen by striking the salt with a hammer in a dark, cold room; flashes of light appear at each blow. A similar fact occurs in chopping green, frozen wood in very cold weather, the axe seeming to strike phosphorescent fire from the wood. In these and similar examples it will be noticed that light is found *to exist, radiating from the sun; then as absorbed and latent in sugar, ocean water,* diamonds, spars, &c., *then as phosphorescent light* surrounding bodies, analogous to the sun's photosphere; and in the process of being emitted from phosphorescent bodies we perceive that light moves *analogous to the emission of heat from metals—a mode termed conduction.* A better example of both heat and light moving by conduction, is presented by an electric current, as the current embodies and carries both of these agents. The fifth mode in which light exists, that of convection, will be noticed in another place.

These examples show clearly that light, in its components of illuminating rays and heating rays, exists in five distinct conditions and similar facts may be traced by following the rearranging rays. *The trinity group are radiated, conducted, become latent and phosphorescent, both as a group and singly.* The rearranging rays of the sun's beam are the rays that cause the skin to darken, fruit to ripen, leaves to color, and colors to fade; *this* entire *group of rays and the material substance* that enters into the structures of organic life *are mutual counterparts.* The agents of the sun's group retain, under all circumstances, and in all conditions, the properties of gathering information, enrolling knowledge and embodying actual force, so that when absorbed into material substance, *this group of agents imparts to the substance,* by their presence, *intelligent tendencies and executive force.*

The influence and remarkable change which heat imparts to oxide of hydrogen when absorbed and remaining latent, rendering an otherwise worthless substance one of the most useful and important of compounds to organic life has already been adverted to, and as we turn to the air we breathe and examine this essential substance we shall find still more prominent and startling phenomena of vital processes due to the absorption and influence of another member of the sun's rays—*the absorption of rearranging force in atmospheric oxygen.*

The doctrine that the primary elements of matter are absolutely endowed with certain inherent and persistent properties has been so persistently taught, *asserted, and reasserted* so often as to have induced almost universal belief. This doctrine of inherent properties has led to the fallacious hypothesis that the phenomena of both the organic and inorganic world are spontaneous results of inherent *material properties.* For one hundred years or more since the chemist, Lavoiser, suggested that the lungs of an animal consumed oxygen like a furnace, the vital processes of human life have been regarded as chemical processes. There have been *imaginary* combustions of carbonaceous matter in all parts of the human system, and a chemical solution of nutriment in the stomach, like the conversion of ores into salts by acids in chemical factories. Yet the facts which have now accumulated respecting the allotropic states of matter and the allotropic states of force, *enables us to assert*, without the slightest conflict with any established fact, that the ultimate fibres and structures of life are not composed of a single chemical compound, nor constructed by a single chemical process. These broad statements, or assertions, conflict with cart loads of theoretical explanations of vital processes by chemical theories, but not with one solitary and positive established fact.

Substance of chemical construction found within organic structures is confined entirely to channels of distribution and finally eliminated, instead of being built into living forms of life. There is probably no one substance that has contributed so much to establish these fallacious chemical theories of organic structure and organic processes as the material element of oxygen; no substance that has been so loaded down with imaginary functions and capacities— it is the sheet anchor of materialism, and the corner-stone of imaginary material powers and capacities.

Oxygen forms one-fifth of the atmosphere, one-ninth of water, and about one-sixth of the known earth, existing in the three states of solid, fluid and gaseous matter. It exists in five allotropic conditions, the two best known being ozone and oxygen, as it exists in the atmosphere we breathe. As ozone, oxygen is highly charged with chemical affinity, while as combined with nitrogen of the atmosphere it is totally devoid of chemical affinity; and it might be here stated that the allotropic conditions of the material elements are founded on their relation and combinations with imponderable agents. The agents of heat, of affinity, of electricity, of magnetism, may each or either exist in material substance, and may each and either become superceded and displaced by one or more of the other agents, such a supercedence and displacement constituting an allotropic change—a change that may be more of or less permanent. In the red variety of phosphorus already referred to, chemical affinity is largely displaced, while the fluid variety has absorbed a surplus amount of heat. Oxygen, in the condition of ozone, is highly charged with chemical affinity, and if a strip of silver foil, or a silver coin be placed in a bottle of oxygen in this condition, it is quickly oxidized and converted into dross, while it might remain in atmospheric oxygen forever without oxidation or becoming the least tarnished.

Atmospheric oxygen and nitrogen do not show the slightest tendency for chemical union; their union by chemical affinity would develop heat and form nitric acid or nitric oxides.

Each of the other imponderable agents have their own special work, and have their own modes of development, each never interfering with the modes of development or with the duties of the other. Chemical affinity unites material elements and forms them into solid crystalline combinations in definite proportions by weight; but in the ultimate parts of organic structures these definite chemical compounds are never found—the so-called phosphates of lime, of which bones are said to be composed, are formed by the processes that find them. Bones are in constant process of change from the embryo to the grave; in early life they are gelatinous and flexible, the outer parts comparatively hard, the inner tender, the whole gradually growing hard, brittle and of different composition through all the periods of life—a feature that never obtains with chemical compounds.

Organic compounds, united by organic affinity, are non-crystalline in structure, not solid but porous. This porous condition of combinations for organic structures enables the interchanging processes of life to proceed continuously.

The rocks and solids of the earth are joined by chemical affinity, the gaseous envelop by organic affinity, while water is an example in which both organic and inorganic, or chemical affinity is represented. Its elements of oxygen and hydrogen are joined by inorganic affinity, and previous to its absorption of heat is crystalline: after absorbing heat it also absorbs and retains organic affinity by which it forms solutions and absorbs gases.

Carbon and each and all of the material elements that enter into the construction of living structures are as devoid of chemical affinity as gold is of magnetism. The absence of chemical affinity may be traced in other forms of carbon and carbonaceous matter besides structures of life. Tons of diamonds might be placed in furnaces, yet they would be found so totally free from chemical affinity that they would not burn. Even black lead is so free from this affinity that it requires a white heat to throw it into a condition to absorb sufficient chemical affinity to induce combination with oxygen. Phosphorous, which in one condition melts and takes fire in in warm hands, in another allotropic condition has been exposed to a concentrated heat sufficient to melt cast iron in a few seconds, yet failed to even melt. The allotropic states of the material elements, and the allotropic states of the forces that enters into the construction of organic life will be discussed again in their relation to chemical force. Sufficient has probably been already said to fix the fact within the mind of the reader that imponderable agents, as forces in sunbeams, can and do exist independent of material substratum, then become absorbed by the substance of matter, imparting to matter allotropic changes, and sustaining allotropic changes themselves by which both forces and substance becomes adapted to organic life.

ORIGIN, USES AND DESTINY OF NERVE FORCE.

It is an established, and an unquestioned fact, that the food we eat, the water we drink, and the air we breathe, undergo certain interchanging rearrangements of their substance within the animal body, and that from these rearranging processes the vital forces of the organism emerge and become manifest. The view herein presented asserts that these vital forces, including the force that traverses the animal nerves, were actually absorbed from sunbeams by the substance of food, drink and air, and are retained by these substances in a dormant state; and that the presence of these agents, retained within the substance of food, drink, and air, imparts to their substance the actual properties that adapt them to animal life; that during the molecular interchangements of the substance of food, drink and air, these forces and agents are developed from their dormant states, and produce their several manifestations of functions and processes; that within the lungs, heart, stomach, and each of what are termed vital organs, the sunbeams group of forces becomes released and reappears as nerve force, substantially as electricity becomes released in battery interchanges.

From what has now been said of nerve force we can trace in a more connected manner the origin, uses and destiny of this queen of imponderables. Discarding for the present all conditions or manifestations of the force, except those presented by animal nerves, and reverting only to those conditions and processes that develop actual nerve phenomena, we find it primarily dependent on the absorption of oxygen from the atmosphere by the animal system—air breathing animals all being provided with nerves and manifesting nerve force.

The atmosphere, as already noticed, is a combination of oxygen and nitrogen, united by an affinity that belongs to organic matter, and to the organic world—an affinity that is as distinct and different from the affinity that unites the substances of water, and stones, and earth, and inorganic matter, as the two kinds of electric attractions—voltaic and surface—dynamic and static are different. On reaching the lungs this organic combination of the two gases becomes decomposed, oxygen being separated from its gaseous mate and uniting with the blood—blood offering a stronger attraction for this element than is offered by nitrogen. In this transfer of oxygen from one compound to another there is a process that is very analogous to a transfer of oxygen from one compound to another in voltaic batteries that produces, or develops, a voltaic current, with this important difference: the combination that becomes decomposed in a voltaic battery, to produce a voltaic current, *is a chemical combination*—a compound of elements that are held by *chemical affinity*—while the combination that is decomposed in the lungs is a compound of elements that are held and united by *organic affinity*. In the battery, water and acids are decomposed. These combination of elements are chemical compounds, joined by laws of definite proportion; while the atmosphere is a combination that is not a chemical combination, and is not held by chemical law. In the first union of oxygen with the blood it does not unite by chemical affinity, or force, but by the same kind of affinity which joined it with nitrogen—an affinity that forms solutions, as when water takes up oxygen or other gases, or when it forms solutions, as of salt, sugar or milk. This distinction between the two kinds of affinity is essential; for what is insisted on here, is, that the *nerve current* sustains the same relation *to organic combinations*, and *organic interchanges*, that the *voltaic current* does to *chemical compounds and chemical interchanges;* that the subtile nerve current is developed the instant that oxygen separates from nitrogen and unites with blood, in the same sense that the subtile electric current is developed the instant that oxygen separates from water and unites with zinc, in a common battery.

In the lungs there is a vast network of nerves, distributed over their entire surface, that receive the subtile agent as fast as developed, and distributes it to other points with which they connect. The network within the lungs serves to collect the agent, just as a network of metal wires might serve to collect an electric current and distribute it to other points. In other words, it is understood that zinc, within a galvanic battery solution, offers a greater inorganic affinity for the combined oxygen of water than this element has for its hydrogen; and this stronger affinity leads to the formation of oxide of zinc, the transfer of oxygen developing the electric force and current; and so in the development of the nerve force, and current, the organized blood offers a stronger *organic affinity* for atmospheric oxygen than nitrogen has for this element; the change from nitrogen to animal blood develops the nerve force and nerve current. This interchange of substance, for the development of the messenger that traverses the nerves of animal life, is but one of a vast series of interchanges of elements, held by organic affinity, that produces, or develops nerve force. The same principle obtains in the development of nerve currents, that obtains in the development of voltaic currents—that is, an almost endless variety of different substances may be substituted, one for the other, but all involving the same general principle of interchange of substance presented by the transfer of oxygen from water to form oxide of zinc. There is the decomposition of one compound to compose another similar compound, the one noticed being from oxide of hydrogen to oxide of zinc, to develop electricity in the battery, and from atmospheric oxygen to animal blood, to develop the nerve force, each of these forces exist in dormant states in their respective compounds, and are developed from the inactive, to the active state.

In a Smee, or common battery, in addition to the separation of oxygen from water, there also occurs separation of oxygen from sulphur, of the sulphuric acid, to unite with the zinc; also, separation of sulphuric acid from oxide of hydrogen to unite with oxide of zinc, each of these transfers, and interchanges, assisting to develop the electric force. And so of the successive processes, which food undergoes, in each of the several stages of digestion. There is a separation—a detaching of a part of the substance of the food, and a re-combination of the detached part with some part of the organism. In the battery there is a decomposition, and a separation of an element, or chemical combination, as of oxygen or sulphuric acid from water, and its combination anew with another substance; and the character of the substance decomposed is reproduced in the compound formed; it is substance held and united by chemical force that is decomposed, and it is substance held and united by chemical force that is again formed—chemical substance decomposed, and chemical substance again composed—while in the animal system, that which is decomposed and detached from combination is substance *held by organic affinity*, and in a different allotropic state from battery fluids.

It is oxygen, held with nitrogen by the force of organic affinity, that is breathed by land animals, and it is oxygen, dissolved and absorbed, and held by the same organic affinity in water, that is breathed by water animals; and it is only oxygen that is conjoined by this affinity, that becomes detached and separated for new unions within the animal system. It is an essential fact that the combinations of oxygen, held by this kind of affinity, are the only ones that are adapted to the process of breathing, or from which oxygen can be separated and detached for animal use. It is not simply that oxygen of the atmosphere exists as a gas, that adapts it to the process of breathing—for it is breathed by water animals, as dissolved in water, and in a state in which it cannot be said to exist as a gas, while ozone does exist as oxygen gas, but in a state or condition that is fatal to animal life when breathed. In the state of ozone, oxygen is charged with chemical affinity, in the same sense that steel can be charged with magnetism; while atmospheric oxygen is charged by the sun's rays with organic affinity, and it is this organic affinity, with its fellow members of the sunbeam, that is transferred and developed, to exist as nerve force. Neither the lungs, nor any other animal organ has the power, or exercises the function, of separating oxygen from oxide of hydrogen, or any other *chemical combination* of oxygen for animal use; it is a substance essential to life, but it is essential that it shall be charged with *organic affinity*. Water is a chemical combination, but it is totally devoid of *free* chemical affinity, being entirely neutral and also charged with both absolute heat and organic affinity—its organic affinity being the affinity that enables it to dissolve oxygen and form other solutions.

Ozone is oxygen gas charged with chemical affinity; atmospheric oxygen is oxygen gas charged with *organic affinity*—these being allotropic conditions and states of the same material substance—ozone being adapted to the inorganic world, atmospheric oxygen being adapted to the organic world—the combinations, decomposition, and combination of ozone compounds developing electric currents, the decomposition and combination of atmospheric, organic or oxygen compounds developing the nerve force.

It is also an essential and positive fact that when oxygen is absorbed and dissolved in the blood, it shall be united by the same organic affinity that dissolves it in water for animal use; essential that it shall not join the circulating fluid and form chemical compounds, but be carried along as an organic partner.

It will thus be seen that the origin of the two forces (nerve force and electric force) are somewhat analogous in general principle of development, but distinct in the fact of being developed from different allotropic states of matter, and by interchanges of compounds, held by different classes of affinity. Then as it is a positive fact that the allotropic states of matter are due to the fact that the substance has actually absorbed and retained different imponderable agents, we can readily understand why substance, as oxygen, in one allotropic state, shall hold for development electric force, and when in another allotropic state shall hold for development nerve force.

The interchange of oxygen from one substance to another within the lungs, for the development of nerve force, is but one of a vast series of interchanges of substance, held by the same class of affinity, that takes place for its further development in other organs and parts of the animal system. We catch a partial glimpse of the general principle, as we pass in review, first the action of atmospheric oxygen and its interchange for developing nerve force in land animals and then compare it with the process, as it takes place in the gills of water animals. Oxygen held in water, and that becomes separated by the gills of fish, is no more perceptible than the affinity that holds it dissolved in the water; yet the *element* and the *rearranging force*, which the element holds, are both transferred from the current of water, and *both* contribute a share in the vital process.

It is not only the fact that it is oxygen that is held in the water for transfer for the animal's use, but it is this fact, in conjunction with the other fact, that it holds condensed sunlight within its actual substance, and that these condensed sunbeams determines the character of the force developed.

Similar transfers of substance, holding condensed sunbeams in their substance, takes place in the processes of digestion, and in those of assimilation. The processes of separating oxygen from water by the gills of fish, is a process that separates an invisible substance and an invisible force from the fluid water; and in the process of digestion a similar fact occurs of the invisible force of condensed sunbeams, separating in the process of transfer, *and is taken up by the nerves of the organ as nerve force.*

If, instead of following out in detail the development of nerve force in the organs and processes of digestion, we first turn to some of the uses of nerve force in the phenomena of animal life, we shall find other analogies between the uses of an electric current and nerve force, and also the continuation of the principle of allotropic differences. In the simplest of batteries, as of a Smee, there is not only a separation of oxygen from water and its combination with zinc, but there is also a detaching and discharge of hydrogen from the platina plate. This discharge of hydrogen, and the combination of oxygen, are simultaneous facts in the development of the electric current; and we find a corresponding fact within the lungs; there is a detaching and discharge of carbonic acid gas from the lungs and blood, as oxygen is absorbed. Hydrogen gas is only eliminated when the plates are electrically conjoined, the discharge being determined by the same facts that determine the current. The same principle also obtains in the elimination of carbonic acid gas from the blood; that is, *the gas is eliminated by the nerve current.*

The celebrated French physiologist, Bichat, divided the study of the nervous system into two classes, nerves of animal life, and nerves of organic life. The nerves of organic life lead from the vital organs, lungs, heart, liver, stomach, &c., to a large nerve extending from the base of the brain, consisting of two cords, and running the entire length of the back bone, termed the great sympathetic nerve. These cords are as large as a pipe stem, of a greyish red color, and lie close to the vertabræ, one on the left, the other on the right side. This sympathetic nerve can be best understood by regarding it as *a nerve switch* arrangement for collecting and distributing the nerve force, in the same sense that an *electrical switch* serves to collect and distribute electricity from many batteries, and distribute or send it to different points for use.

As already noticed, the transfer of atmospheric oxygen from its associated nitrogen, and its absorption by the blood, determines the development of nerve force. This, however, is to be understood as but a degree, or part of the sum total that is developed, or produced, for carrying out the processes and functions of organic life. But what is produced within the lungs is produced in the same sense as electricity is produced within a Smee battery by the transfer of oxygen from hydrogen of the acid solution of zinc. *The nerve force* developed within the lungs, is carried by a nerve line that connects with the sympathetic nerve, substantially as we may suppose an electric current may be carried along a metal wire to an electric switch. At the point where the nerve from the lungs, or other vital organ, connects with the switch, the cords are enlarged, and the enlargements are termed ganglia. Each of the switch cords, at the gangliatic enlargements, receive nerves from two or more organs that, like the lungs, serve to develop nerve force. From the base of the brain to the lower end of the back bone, there are twenty-nine enlargements of each of the switch cords, and each enlargement may be understood as receiving and distributing a distinct degree and grade of nerve force. In other words, if we consider the force developed within the lungs as corresponding with a Smee, or one kind of battery, the heart and blood vessels, stomach, liver, kidneys and other vital organs would correspond with other classes of batteries, and the vital group of organs, would consist of twenty-nine batteries of nerve force, each of these twenty-nine batteries sending two or more lines of nerves to the main switch.

When several electrical batteries are grouped and connected together, the force eliminated from the combination acquires a character known to electricians as intensity, and the intense force becomes competent to produce effects that a single battery is incompetent to produce. Two or more batteries combined will decompose compounds that a single battery is incompetent to decompose. The combination of different numbers of batteries enables us to grade chemical force and detach, for example, either peroxides, protoxides, suboxides, or metals from their chemical combinations. A similar fact obtains with nerve force in the animal system, it is graded so as, for example, to eliminate from the lungs carbonic acid gas, from the kidneys and bladder, water and uric acid salts, while from the stomach and digestive canal, refuse is eliminated to pass from the rectum. Each of the vital organs of the human system *that serves as a battery to develop nerve force* is not only provided with nerves to receive and carry the nerve force to the main switch, but each organ receives a nerve line from the switch that brings a graded degree of nerve force for use. The line of nerves that return from the switch to the lungs carries just that degree of nerve force competent to discharge or disengage carbonic acid gas.

By this view of the nerves of organic life, each battery of nerve force is provided with *two systems of nerves*, one system gathering the force which the organ produces, the other producing a certain result of molecular structure. One system gathers the force, the other uses the force in arranging the anatomical elements in the living structure. The ganglia that receives the nerves from the lungs, also receives nerves from the heart and stomach, which meet in the switch over the vertabræ or in the neck. The stomach also sends nerves to ganglia, meeting nerves in the switch opposite the waist from other organs. Similar arrangements and combinations of nerves from different organs, and their connections at different parts of *the switch*, is true of all of the ganglia, and *is* evidently an automatic arrangement for using the nerve force for carrying out the plan of the organism. It is somewhat analogous to selecting a few from a group of galvanic batteries and using the electric force from the few for special purposes. Two or more combinations of galvanic batteries imparts to the current what is known to electricians, as intensity; and from a chemical point of view, different degrees of intensity correspond with modifications of inorganic affinity. And so from an organic point of view, these combinations of nerves in ganglia determine grades and modifications of organic affinity; for the nerves that lead from the different gangliatic centres, effect different molecular changes at their different points in the organism.

It is also understood by physiologists that organs which send nerves to any particular point of the sympathetic nerve, have closer relations of sympathy with each of these organs than they have with other organs that connect in ganglia at other parts of the nerve. The part of the sympathetic nerve that receives a bundle of nerves is regarded as a nerve center, in the same sense that the brain is regarded as a nerve center; and the force that reached one of these nerve centers, on the switch cord, is distributed, or dispatched, from this center without going to the brain for orders. The successive changes which food undergoes, first in the stomach, then entering the circulation and going to the heart, then impelled to the lungs, brings these three organs into very close sympathy in development and use of the force developed. Then again the stomach by its gastric fluid, the liver by its bile, the pancreas and spleen by their respective contributions toward digestion, are in close sympathy, and their nerves are also grouped in ganglia on the switch.

In other words, the force developed in different organs that is distributed by nerves that are collected in groups in the sympathetic nerve cords, consists of twenty-nine modifications of organic affinity; and these twenty-nine modifications are *automatically developed* from condensed sunbeams, *automatically received* by the *sympathetic switch* cords, and *automatically distributed* to produce the phenomena of growth. Each of the vital organs contributes a degree of nerve force to some part of this automatic switch, and each organ receives in return a degree of force that assists to purify the material substance that is being built into the structure. One class of this system of nerves serves to collect the force, the other class serves to use the force in reconstructing the organic molecules and anatomical elements, and building them into the various parts of the living structure.

Each of these phenomena of purification, and of organizing the purified molecules into living forms, are known to be largely dependent on the forces that traverse the nerves; for if the nerve that regulates these effects be severed the performance of the function ceases.

The substance of food consists of parts that are to be built into the living structure, and also of parts that are to be eliminated as worthless refuse. The parts that are eliminated have assisted the other parts to hold the several forces in dormant states of inaction; and from the parts eliminated as refuse, allotropic states of forces, imparted to them by the sun's rays, become changed to the active conditions, and these agents are taken up by the nerves as nerve force. Refuse from the bladder, alimentary canal, and lungs have lost their absorbed sunbeams, and the lost agents *are transferred to the nerves*—transferred in the same sense that electricity, transferred by battery interchanges, becomes transferred to battery plates to traverse connecting wires.

The production of nerve force and its use are so mutually related that it is impossible to study the origin without at the same time studying the use made of the force by the living processes and structure. In this respect nerve force is like voltaic force, the study of the various uses of the force assist in explaining its modes of production. In tracing the origin of a part of the force produced in the human system, to interchanging processes that take place in the lungs, we find a close analogy, in principle, with the origin, and one mode of producing electric force; and in tracing the uses of nerve force within the animal system, we shall find other close analogies with the uses of electric force. But in all their analogies of development, and in all their analogies of use, there is, and remains, the fundamental allotropic difference of organic and inorganic, living and dead. The interchanges in the organic that develop nerve force are wholly and entirely *non-chemical* interchanges; the interchanges in the inorganic that develop voltaic force are wholly and entirely *chemical* interchanges. Within the voltaic batteries, and within the galvanizing troughs to which the currents are directed, what occurs is decomposition of compounds into their elements, and the formation of similar new compounds from these elements; what occurs within the living animal *is separation of organic molecules and their reorganization in new forms*, brought about by the circuit of nerve force developed by rearrangements of nutriment substance, and traversing this automatic arrangement and circuit of nerve lines.

By virtue of this principle of separation of substance held by organic affinity, and its reorganization, *the mouth, by means of its apparatus of salivary glands, teeth and nerves, sustains a position as one of the organs or nerve batteries for developing nerve force.* By means of its disintegrating apparatus, its saliva is brought in contact with minute parts of nutritious substance, starting interchanges between nutriment substance and saliva, thus releasing rearranging force that is taken up by the surrounding nerves. There is not only the sensation of taste from this process, but there is also rearrangement of the nutriment substance by the actual rearranging force previously absorbed from the sun's rays by the substance. The rearranging process, thus started in the mouth as the first enlargement of the alimentary canal, is continued in a graded series of changes, each vital organ retaining the substance surrounded by its network of nerves during the process of rearranging the nutritious substance—one grade of the series—the stomach, duodenum, intestines, lymphatics, arteries, veins, lungs and others, all serving as nerve batteries, that receive by their system of nerves nerve force that is sent to the automatic switch.

The process of digestion is well known to be mainly under the influence and control of the system of nerves that invest and twine around the digestive apparatus. The digestive apparatus is not only provided with nerves to regulate the process of digestion, but it is also provided with a system of nerves that gather the nerve force produced or developed by the digestive process. In the language of Gray's Descriptive Anatomy, nerves distributed from the sympathetic cords, " Have a remarkable tendency to form intricate plexuses, which encircle the blood vessels and are conducted by them to the viscera." This tendency of the nerves from the sympathetic, to encircle the blood vessels and viscera with network of nerves, by the view here presented, accomplishes the double purpose of gathering force set free by the rearranging processes that take place within the viscera, molecular interchangings and rearrangements of nutriment occurring in successive stages from its first entrance within the mouth until it is woven into tissues and built into the completed structures.

As already noticed, the two classes of nerves that encircle the viscera follow the arteries and veins throughout their entire circuit; throughout the entire arterial course one class of nerves gathers force, and *throughout the circuit of the veins, force is despatched to eliminate carbonic acid gas from the lungs.*

This power and property of the return line of nerves from this switch to eliminate, set free or deposit organic molecules, is analogous to the power and property of the return line of an electric circuit to eliminate, set free, or deposit from battery and galvanizing fluids, inorganic molecules and elements. In a Smee battery, hydrogen is eliminated from the platina plate, or return line, but in a sulphate of copper battery, the hydrogen, instead of being eliminated as gas, unites with oxygen of the copper oxide, reproducing water which remains in the solution. The hydrogen is got rid of by combining it with substance within the battery; the water thus formed becomes a useful product, and, by the union of its elements, force is also set free, augmenting the battery force. A somewhat analogous mode of getting rid of carbonic acid gas, eliminated by animals, occurs in some classes of low orders. Corals, clams, snails and many shell animals, instead of eliminating carbonic acid gas, like birds and quadrupeds, concentrate the gas in the earthy matter of their shells. They withdraw oxygen, condensed in water, as fish do; but the return force, instead of eliminating the gas induces its combination with earthy matter to form their shells. The mouths of snails have been pasted shut for years without smothering them—an effect that would have occurred but for the mode provided for getting rid of the deleterious acid.

The power and function of the nerve currents from different ganglia, to eliminate, or to deposit anatomical elements—to eliminate saliva from the salivary glands; bile from the liver; uric acid solutions from the kidneys, or to deposit in structure, bones, muscles and bonanzas, built into the forms of life—is analogous to the uses that may be derived from different groups of battery currents to separate, set free, or deposit different chemical products.

The arrangement and combination of nerves that are distributed from the sympathetic to viscera, again form ganglia on their continued lines from viscera to accessory, or minor organs, and also again spread out in plexuses, and twine around these minor organs. These secondary series of ganglia, like those on the sympathetic cords, are collections of nerves that are thus evidently arranged to subserve the same purpose as the primary groups of nerves on the main switch; that is, they are automatic arrangements for the distribution of nerve currents. The entire system of nerves from the sympathetic are nerves of involuntary life; the force which they distribute, *automatically carries out a predestined plan of warming and guiding the material molecules, and building them into the forms and structures of living castles.* Through the influence of the agent that traverses the lines from the sympathetic, *the rearrangement and the reorganization of organic matter is effected;* and neither the brain, nor the mind, by taking thought can add one line to its height, or remodel its form, complexion, color or composition.

The automatic arrangement of different lines of nerves to different organs and parts of the organization by which, in some cases, continuous processes as of secretion take place, and in other processes, discontinuous and periodication occurs, as in wakefulness and sleep, in breathing, beating of the heart, evacual discharges from the bladder, rectum and liver, all point to the fact that some mode *of regulating the supply of current to different organs* exists in the sympathetic system of nerves—some mode by which it may be, and is, automatically increased and diminished to carry out the plan of organization.

The voluntary system of nerves reveals the fact that through nerves which control voluntary muscular actions the nerve current is sent at the option of volition. The sympathetic system of nerves, for example, distributed to the hand, regulates the separation, assimilation and deposition of its substance, in all its various forms and modifications, while the voluntary system of its nerves regulate and control its movements in grasping, lifting, writing, &c., &c.—the one system sending optional and voluntary currents, the other revealing automatic and involuntary currents of nerve force. Each system, however, reveals the fact that nerve currents, instead of being uniformly equal and continuous, are periodically increased and diminished, imparting greater or less degrees of force at various times. In dyspepsia, fright and fevers, we see digestion suspended for hours, days or weeks, while in locomotion and muscular work we see voluntary states of intensity, interspersed with voluntary states of rest and inaction, all of which reveal the fact that the nerve lines are provided with arrangements and modes of regulating the force and messenger current sent through the nerve paths.

The manner in which this is brought about, involves the study of nerves, their mode of connection, conduction and modes of influence, both on contiguous nerves and on other forms of matter. This feature or mode of regulating the nerve current, is here passed and reserved for future discussion.

Voltaic batteries, not only develop chemical force which can be transferred to plating baths and galvanizing troughs, to rearrange the elements of chemical compounds, but the current, in its circuit, also at other points, develops both heat and magnetism, and the nerve current shows not only rearranging force in rearranging organic compounds, but it also *reveals the properties of developing and manifesting heat, muscular force and intelligence.*

The discussion of the analogies between the modes of developing magnetism by voltaic currents, and the development of muscular force by nerve currents, is reserved for separate discussion; but there is one feature of their relations that brings muscular force into very intimate relation with the vital processes controlled by the sympathetic system of nerves. This feature is, that nerves which influence and control involuntary muscular actions, such as beating of the heart, expansion of the lungs, peristaltic motions of the bowels, &c , &c , are associated with nerves from the sympathetic system in very close relations, but not joined by direct union. Nerves from the two systems meet in the ganglia of the sympathetic, and are enveloped in the same sheaths reaching the heart, stomach and other vital organs, in this close proximity ; but the microscope has revealed the fact that notwithstanding this close proximity, the nerve fibres from each system retains its individuality and never merges into that of the other.

Each line of nerves, whether from the sympathetic cords or from the spinal marrow, are continuous on themselves as distinct fibres, often lying side by side, but insulated, each from the other. Arranged in this way the two systems of nerves exert a mutual influence over each other; each contributes to the harmonious action of the other, but also forms an arrangement by which disorders to one system works injuries to the other system; each contributes towards the functions of life, but at the same time each may interfere with the other's functions. When the celebrated phisiologist, Flourens, discovered the extremely dangerous sensitiveness of that part of the spinal marrow where it joins the brain, he imagined that he had discovered the seat of life. The prick, or penetration by the finest needle, of the spinal cord at this point will instantly destroy any animal. The nerves that control the expansion of the lungs, connect with the cord at this point, and, if severed, or their function of conduction is destroyed, the lungs can neither absorb oxygen nor expel carbonic acid—the stoppage of either function being fatal to animal life.

The peristaltic motion of the bowels, and the rotating movement of the stomach, are brought about by the nerves of animal life, while the interchanging processes of the nutriment matter within is governed by the automatic system, yet each is essential to the other, and mutually governing each other by the principle of induction.

In the phenomena of human life the action and principle of induction is a very broad and important one, both in the explanation and its use in both normal and abnormal phenomena.

In discussing the influence and power of nerve force the phenomena of *induction* will be referred to frequently as the force which, in one condition traverses the animal nerves, and in other conditions exerts equally important influences *by other modes of propagation*.

Physiologists have traced and discovered that all nerves which influence the shortening and lengthening of muscles and muscular fibres, extend from the muscles and fibres to the spinal marrow; and have also determined that all bodily motions, both voluntary and involuntary, are brought about by these primary motions of elongating and contracting muscular fibres. In other words, there is not only a system of nerves from the heart, stomach and other vital organs, that are arranged in the ganglia of the sympathetic switch to automatically carry on the processes of life, to weave, and mould, and build the material substance into living structures through the processes of growth, but there is also another system of nerves extending from these vital organs to the sympathetic switch; but instead of stopping in the automatic switch, these nerves pass through the ganglia and connect with the spinal marrow. These two classes of nerves belong to Bichat's division of nerves of animal and of organic life. The animal nerves govern bodily motions of muscular fibre, the organic division govern molecules for growth. The two systems extend from the vital organs to the ganglia together, and often enveloped in a tubular tissue.

From an electrical point of view, these nerves lying side by side, but not joined by substance, to conduct the nerve current, are analogous to two wires lying side by side, but insulated from each other. When one of these wires forms an electric circuit with a voltaic battery, the other wire is influenced by the current, and manifests, what is known to electricians, as an induction current. The induced electric current is a discontinuous, or pulsating current, being increased and diminished by augmenting, or retarding the primary current.

This brief glance at the anatomical arrangements of nerves for muscular work in the spinal cord, and of the arrangement of nerves for molecular work in the automatic switch, is not for the purpose of discussing anatomical facts, but for the purpose of fixing clearly in the mind the *prominent fact that nerve force, instead of being " brain work," or manufactured by the brain, is developed at the other end of the nerves;* that instead of being primarily derived and sent from the brain to the lungs, stomach and vital organs, *it is developed within these organs from condensed sunbeams* that have been absorbed and retained in the substance of the food we eat, the water we drink, and the air we breathe.

There are low classes of animals in which no brain, or anything that would answer for a brain, has been discovered; but they have nerves and manifest nerve powers. They have arrangements of a few ganglia corresponding with the automatic or sympathetic nerve switch, and a nerve that corresponds with the spinal cord that governs their motions; one class regulating the function of growth, the other that of locomotion.

In the phenomena of human life, disordered states of one part of the organism become manifest in disordered functions in other parts of the body. A severe burn on the body may bring on chronic diarrhea; indigestion, or a sour stomach, may bring on severe headache; these, and similar interferences, that often develop abnormal functions, reveal capacities of the force that traverses the nerves to transmit abnormal, as well as normal effects, just as undesirable results obtain in galvanizing from deranged battery action.

One of the first things noticed by students is the analogy between the functions of vegetable leaves, and the lungs of animal life. Leaves breathe carbonic acid gas; lungs oxygen gas. *The absorption of carbonic acid imparts vigor of growth to the vegetable kingdom;* its expulsion is essential to impart energy to animals. The early investigators of organic processes inferred that carbonic acid, imbibed by vegetation, was decomposed, and its carbon appropriated by the growing plant. Other investigators traced the absorption of oxygen within the lungs, and the expulsion of carbonic acid, and inferred a combustion of carbon. Volumes have been written on this mutual relation of vegetable leaves to animal lungs, and for generations, instead of its decomposition and its recomposition being stated as an inference, this decomposition of the acid, and its recomposition within the animal, have been stated as positive facts; and to question its accuracy will, to some, seem almost a sacrilege; yet the evidence which has accumulated from the labors of later investigators, almost irresistibly shows that the leaves of vegetation do not decompose carbonic acid, nor do the processes of animal life regenerate this compound.

"The building up of the vegetable, then, is effected by the sun, through the reduction of chemical compounds We eat the vegetable, and we breathe the oxygen of the air; and in our bodies the oxygen which had been lifted from the carbon and hydrogen by the action of the sun, again falls towards them, producing animal heat and developing animal forms." We have, in this extract, such an elegant blending of fact and inference that, without special attention, it would be difficult to tell which is fact and which is inference. A few words, however, will set the matter right. We know that the sun's rays impinge on the growing leaves of vegetation; that the leaves are surrounded by the material substance of carbonic acid; that both the rays and the acid are absorbed, or disappear in the leaves; and the time honored inference is, that the rays have decomposed the acid, and the plants have appropriated its carbon. We also know that oxygen is absorbed by the blood within the lungs, and carbonic acid expelled; and the time-honored inference is that the absorbed oxygen has entered into combination with carbon somewhere within the body, and formed carbonic acid. Neither the decomposition of the acid has been traced, nor its re-composition discovered —both are inferred.

There is no question but that carbonic acid is absorbed by vegetation; and no question but that oxygen is absorbed and the acid expelled from the animal; but there is a question whether carbonic acid is decomposed into its primary chemical elements; there is a question whether these primary elements are chemically recomposed and regenerated within the animal body; there is a question as to how the vital forces of growth are developed and become manifest in growth of vegetation by reason of this disappearance of the sun's rays and carbonic acid within the vegetable leaves; and there is also a question as to how the vital energies of nerve power are developed—emerge and become manifest as a result of the absorption of oxygen and expulsion of carbonic acid from the animal. We also *eat the animal* and derive quite as much nourishment and energy therefrom as from the vegetable.

It is not the design, in this publication, to follow carbonic acid through plant and animal structure, but simply to fix attention to the fact that *nerve force is condensed sunbeams*. It is sufficient here to state that the labors of investigators tend clearly to show that carbonic acid, when it reaches the leaves of vegetation, absorbs light and undergoes an allotropic change of structure, substantially as oxygen does when absorbed by water, or by blood—it becomes a partner of the liquid components of vegetable sap. It can be discovered in all vegetable sap, and in the blood of animals by processes that were formerly supposed, or inferred, to be processes of combustion, or oxidations of carbon. In fermentation, these processes have since been ascertained to be quite different from oxidations—they are the separating and withdrawing a part of the matter of these juices as nutriment by minute living organisms, thus releasing carbonic acid as refuse. The process of fermentation, instead of oxidations, and manufacture of carbonic acid, simply releases and sets it free in its gaseous condition; it is a process in which its absorbed force of concentrated sunbeams are transferred to living matter together with nutritious matter with which it was associated. This, in all probability, is what takes place in the phenomena of human life; the bioplastic forces of life, of which nerve force is one, release carbonic acid by withdrawing its associated material substance, and withdrawing its associated immaterial and condensed sunbeams as nerve force.

The development of heat by chemical combustion is of so impressive a character that the early observers wildly concluded that heat was the prime and only agent in the entire world of phenomena. Among these imaginary combustions, the combustion of carbon and formation of carbonic acid within the animal body, has been the most extensively sounded. Later investigators have discarded the notion that heat of organic combustion is all sufficient to account for organic life; but as the mysteries of galvanic currents were gradually unfolded many thought that the key to organic life was discovered in these currents. These notions are completely exploded, and largely discarded. The combined results of an army of investigators lead to the irresistible conclusion that neither heat from oxidation, nor voltaic currents are essential to organic or animal life; that animal heat and nerve currents are developed by other modes than by the modes of chemical action.

Carbonic acid sustains much the same relation to organic structures and organic forces that water sustains to these structures and forces; they each serve as carriers of force and material. Water is essential to organic life; but it is neither appropriated as water to be worked into the completed structures, nor is it decomposed and its elements appropriated for building structure. It is not essential as so much oxide of hydrogen, but as a substance which readily absorbs immaterial agents and transfers them to living matter; it is the queen of intermediate substance. Neither its molecules nor its component elements form a component part of completed tissues, or structures; but as a solvent and absorbent of molecular bonanzas that are to be woven and built into completed structures of life, and as an absorbent of forces that weave and build these structures, water is pre-eminently the most important substance in organic processes. It is the queen of intermediates; but at the completion of living structure it is discharged from service. It is a chemic compound; and no chemic substance, no compound whose elements are held and united by inorganic affinity, ever receives the crown of life.

The allotropic line between the living and the dead is absolute, and founded primarily on the two classes of affinity.

Water forms nearly three-fourths of the bulk of living things; but it is confined entirely to channels of distribution, and, from the completed structures of life, it is readily separated by evaporation. It is essential to organic process, but totally discarded as useless in the completed structures of living forms.

What has been said of water is equally true of carbonic acid; it is absorbed and becomes an active partner in the components of both vegetable and animal fluids, but not a chemical component; it becomes an active partner by virtue of having absorbed and retained rearranging force. It does not form carbonates by joining these fluids, any more than it does in becoming absorbed in water. It passes through the channels of distribution with water; and, like this neutral chemic combination, it imparts force and strength by virtue of having absorbed force and power, and holding them for ready transfer—parting with forces absorbed from sunbeams, and then resuming its own former state of condensible gas. If carbonic acid was formed by combustion within the lungs or other organ, the heat generated by the union of oxygen and combustibles would become manifest; but it is well ascertained that neither the lungs nor any other organ shows any special rise in temperature from the inferred oxidation. Animal heat is developed by other modes, than by turning oxygen in among the delicate tissues, wildly rampant with chemical energies and affinities, to burn and turn the beauties of life into the foam of inorganic dross. The most delicate search with thermometers, in the lungs, in the heart, blood vessels and capillaries, have failed to detect this inferred oxidation of carbon to form carbonic acid, as a mode of producing animal heat. Animal heat is regulated by the nerves; and nerves have no control over chemical process, as affinities, or products. The changes of carbonic acid in organic process are *allotropic* changes of form, and not of composition.

The atmosphere and the waters of the earth are completely filled with an impalpable dust that impart to the sky an azure blue, and to the waters a darker shade. The atmosphere contains not only oxygen and carbonic acid, as substance that plays so important a part in organic life, but it also contains this azure matter charged with viability—one of the group of vital forces. Viability is the embodiment of germinal force. In contact with living matter this cloud dust, as it has been called, and carbonic acid, unite and condense in the vegetable sap, as one mode of forming protoplasm. This condensation takes place as an effect of allotropic rays—the rays becoming enveloped and permeating the condensed acid with viable force, imparting vital energies to the protoplasm, or bioplasm thus formed, substantially as heat becomes enveloped, permeates, and imparts elasticity to water in steam.

Investigators have followed heat from the sun's rays into the woody fibre of vegetation, traced it from the burning wood to the iron sheets of steam generators, then through the iron plates into the water within where it imparts to the water an entirely new property—that of elasticity—followed the absorbed heat enveloped in the elastic steam, as it is conveyed through hot pipes into the cylinders of steam engines, where this material substance, by virtue of its absorbed immaterial force, becomes a marvel of force and mystery. An analogous fact is true of carbonic acid; the allotropic rays of the sun, with their property of rearranging the molecules of matter, may be traced into carbonic acid, condensing the acid and becoming absorbed with *viable* force—thus imparting to material substance forces and energies that may be traced into the animal kingdom, to reappear as nerve and viable germinal forces.

The condensation of oxygen, and the condensation of carbonic acid, without chemical affinity, are facts well known and established. A given bulk of nitrogen gas will absorb an atmospheric proportion of oxygen gas and the mixed gases immediately shrink into less space than was occupied by the nitrogen alone. Oxygen becomes absorbed by water, and the water by the absorption of the gas, shrinks in volume and occupies less space than previous to its absorption. The same fact of shrinking occurs by the absorption of carbonic acid in water. A similar occurrence may be seen by dropping sugar into water—the water shrinks and occupies less space after dissolving the sugar than when it was pure. These are only a few of the examples that might be noticed, but they all represent power. It requires force to compress water into a less bulk, but its affinity for sugar produces this effect. And so of the gases of oxygen and carbonic acid, it requires power to compress them to a liquid state, but their affinity for water accomplish their condensation. When the primary elements unite by the influence of chemical affinity, heat is superceded and displaced—the chemical affinity remaining in a dormant or latent state and holds the elements in union. When gases, or other substances unite by the allotropic force of organic affinity, producing condensation without eliminating heat, both the heat and re-arranging affinity remain dormant, or latent, *to reappear* as a current traversing animal nerves—a result of interchanging pro cesses which take place within the stomach and other organs. Processes of molecular interchange occurring so largely within the stomach, renders this organ *one of the most important* of animal organs for the development of nerve force.

Having already exceeded the designed limits of this pamphlet, the further discussion of the origin, uses and destiny of nerve force, is deferred for a future publication.

In following the identity of light and nerve force, in further arguments, I propose showing *that condensed sunbeams* are the enrolling agents of animal and vegetable instincts, as well as their intelligence ; and to trace the dividing line between instinct and intelligence from an absolute and allotropic view.

The difference between the animate and the inanimate, between the living and the dead, between, for example, the blood sucking instincts and muscular agility that is embodied in the material substance of a live flea, when compared with the latent chemical inert energies embodied in a grain of sand or a dead flea, is vast and appreciable; and there is no truth which the labors of a host of investigators has more clearly established than the fact that these vast differences are founded on the altered states of matter, due to the displacement therefrom of one class of energies, and superceding them with another class of energies. It is absolutely certain that in organic life, and in organic processes, chemical affinities, chemical processes, and chemical substance, are totally superceded and displaced by a new and distinct group of energies. This supercedence of chemical energies in the principal elements of organic matter has been traced—we have seen that carbon, hydrogen, phosphorus, nitrogen, oxygen and all the primary elements of organic matter are well known in allotropic states devoid of chemical affinity.

Volumes have been written on the inherent property of iron to respond to magnetic attractions. By suspending a bar of iron within a bottle, by a small cord, and bringing a magnet near the bottle, the bar swings towards the magnet by magnetic attraction ; yet, if the iron be thrown into another allotropic condition by combining it with oxygen and sulphuric acid it becomes totally indifferent to the magnet. The atoms of the iron bar may all be suspended within the bottle, some of them being nearer the magnet than in their previous suspension, yet the atoms will be totally indifferent to the magnetic attraction.

The above is a simple illustration and type of the displacement of well known properties caused by allotropic modifications of imponderable agents.

By virtue of similar displacements of inorganic forces and inorganic properties and superceding them with an allotropic class of energies and properties adapted to the organic world, we perceive the working of a principle which carries with it the superceding of organic forces and properties by spiritual forces and agencies adapted to other allotropic states of existence. Huge volumes of inferences of what becomes of man at death—predicated on the so-called inherent properties of matter and brain work—exist, but these inferences and deductions vanish like thick darkness in the morning sun, when brought in conflict with allotropic states of matter and allotropic states of force.

The inference, assumption, assertion and re-assertion, so persistently made during the past generation, that the brain translates, transmutes, and manufactures the hieroglyphics of fact into mental phenomena by virtue of a combustion of its own substance—an oxidation of its components—and that from this burning crash of atoms, emotions, knowledge, memory, and all mental phenomena emerge, are assumptions just about as self-evident as that conscience emerges from the melted lavas of volcanoes, and just about as well established by definite evidence. The assertion that the brain manufactures sound from the vibrations of matter, and vision from the quivering of æther, has led to such a piling of theory upon theory to explain the hypothetical manufacture of all imponderable agents and their groups in living things, as to put to blush the hypothetical powers of pagan idols. The imaginary combustion of brain substance, if it occurs, as an actual fact, from severe mental labor, would leave the human skull, at times, as empty as a coal-bin from which its coal had been burnt; but the absurdity of this conjecture is so intense that no one looks for this kind of evidence to confirm this long taught theory.

The assumption that nerve power and mental states are produced by brain combustion, carries with it the necessary inference that when the brain ceases to exist, mind and nerve force being but temporary thrills of their burning crash of atoms, must also fail to be produced. This inference being founded on non-existent facts of brain combustion, falls into blank oblivion—like inverted shadows from obstructed light—by the luminous facts of allotropic states of matter and force.

The group of forces and components of man's inner self, as well as his outer self, are a group of indestructible elements, gifted with an inborn sense of immortality ; its nerve force, as one member of this group, and as an allotropic condition of light, has the range of the universe—its life period eternity. Inferences and deductions of annihilation, or of an eternal sleep, have no analogies in the world of fact or phenomena.

The facts which have been grouped as allotropic conditions are as familiar as sunshine. The vital group of forces that lie slumbering in the seeds we plant are latent—an allotropic state that, in the phenomena of life, is not less important than the active states with their more prominent manifestations. The inferences and deductions founded on the narrow range of inorganic matter, as revealing total death, are but inverted shadows, instead of actual fact, when applied to other states of matter or force. The inference of an eternal sleep is a spectral illusion, in direct conflict with the universe of fact. The five conditions of matter, and the five conditions of force, reveal an endless circle of change of never ending succession; and during the brief period of man's three score years and ten, the condensed sunbeams that traverse his nerves as nerve force, gathers information and enrolls knowledge, which is there grouped with indestructible force.

The group of forces that compose man's inner self are so completely interlinked that the discussion of any one of them also involves, to a more or less extent, the discussion of the others. These links are especially true of viability and muscular force with nerve force.

In presenting further arguments in support of the identity of light and nerve force, some of the allotropic conditions of viability will be noticed.

In discussing muscular force, from an allotropic point of view, its presence in the inorganic world be traced, and it will also be shown that the influence of nerve force over animal muscles, has its allotropic analogy and correllate condition in the opening of flowers and movement of vegetable fibres, by the allotropic state of nerve force in impinging sunbeams; to which will be added the facts and evidence of tangible experiments.

IDENTITY OF LIGHT AND NERVE FORCE IN ALLOTROPIC CONDITIONS.

———o———

CONTENTS

		PAGE.
I.	Identity of Light and Nerve Force in Allotropic Conditions...............................	5
II.	Allotropic Conditions of Matter and Allotropic Conditions of Force...........................	10
III.	Allotropic Properties of Forces...................	12
IV.	Light Gathers Information, Enrolls Knowledge and Embodies Actual Force...................	13
V.	What Is Light..	23
VI.	Vision...	29
VII.	Sound and Hearing Due to Non-Luminous Conditions of Light	44
VIII.	Undulatory Theories.............................	48
IX.	Smell, Feeling and Taste..........................	68
X.	Allotropic States of Organic Forces............	70
XI.	Allotropic States of Heat—Latent Heat.......	74
XII.	Allotropic States of Organic Forces—Agents Other than Heat................................	77
XIII.	Origin, Uses and Destiny of Nerve Force.....	96

Entered according to Act of Congress, in the year 1879,
By J. CHANDLER,
in the office of the Librarian of Congress, at Washington.
All rights reserved.

Identity of Light and Nerve Force

IN ALLOTROPIC CONDITIONS.

When it became known that LIGHTNING and ELECTRICITY were identical; known that the subtile agent produced by rubbing amber with a woolen cloth was IDENTICAL with the fierce bolt that dashes from cloud to cloud, and that the untamed steed had been caught and harnessed, it gave a new and strong throb to the pulse of human life—bringing into the van of human progress batteries and magnets, plating baths and electric lights, telephones and telegraphs, all impelled by the subtile thunderbolt.

And, when it is known that NERVE FORCE and SUNBEAMS are absolutely IDENTICAL; known that the subtile AGENT that traverses the nerves of the human frame is the same identical agent that traverses space as LIGHT, a new era will dawn on the human mind. And when it is known that NERVE FORCE is no more confined to animal NERVES than electricity is confined to telegraph wires, no more confined to nerve matter than heat is confined to furnaces, *its study becomes intensely fascinating*.

When it becomes known that heat and electricity, and all imponderable agents exist in five distinct allotropic conditions; that NERVE FORCE radiates through space as LIGHT; condenses into and permeates organic structures like animal heat, then reappearing in the third allotropic condition to traverse the nerves by conduction, the knowledge gives a new theme for human thought, and a new force for useful application.

As nerve force, this queen of imponderables, in each and all her allotropic conditions, gathers information, enrolls knowledge, and embodies actual force—imparting to organic life instinct, intelligence and power, the trinity group of knowledge, wisdom, and power to weave, and build, and guide organic structures.

SHOWING ABSOLUTE IDENTITY OF LIGHT AND NERVE FORCE IN SENTIENT AND FORCE QUALITIES,

By J. CHANDLER.

Prepaid, by Mail to Any Address, for 50 Cents,
BY F. M. CHANDLER, TITUSVILLE, PA.

Identity of Light and Nerve Force

IN ALLOTROPIC CONDITIONS.

Further Arguments in Support of the Fact that

LIGHT AND NERVE FORCE ARE IDENTICAL,

WILL BE PUBLISHED,

And the subject discussed in the following sub-divisions:

Origin, Uses and Destiny of Nerve Force, continued.

Condensed Sunbeams the Enrolling Agent of Instinct.

Dividing Line between Human Intelligence and Instincts, from an Allotropic Point of View.

Muscular Force and its Allotropic Conditions.

Viability, (or the Germinal Force,) and its Allotropic Condition.

Mind Reading and Spiritual Manifestations, from an Allotropic View.

Allotropic Relief for Nervous Exhaustion.

www.ingramcontent.com/pod-product-compliance
Lightning Source LLC
Chambersburg PA
CBHW030903170426
43193CB00009BA/721